CW01649595

Handbook

Growing & Managing Acacias in South Central Africa

This book is dedicated to the memory of Dr Richard Barnes who died in May 2004 and deserves recognition as the prime mover and co-ordinator of invaluable research on African acacias.

Richard Delano Barnes was born in Zimbabwe in 1934, and following the forestry undergraduate course at the University of Oxford became successively district forestry officer, tree breeding research officer and director of the Forest Research Centre, Zimbabwe. In 1980 he joined the Commonwealth Forestry Institute where for some years he led the Forest Genetics Research Group.

In the mid 1980s, the international approach to forestry research changed focus from commercial plantation forestry to tree-based rural development. Richard's personal research moved from the genetic improvement of tropical pines for plantations to the taxonomy, ecology and variation of *Acacia* species appropriate for rural development. He embraced this change with characteristic enthusiasm, thoroughness and determination, becoming one of the world's leaders in understanding the biology and applications of the group.

A large number of books and scientific papers have already appeared with Dr Barnes as the major author and they will long remain as standard works and as testimonies to his own research, vision, leadership and commitment to appropriate African development.

Acknowledgments

This book was compiled by Newton Spicer, Richard Barnes, and Jonathan Timberlake, and illustrated by Joseph Mutasa and Rosemary Wise, with additional written contributions from:

Faith Chikomo
Irene Kadzere
Crispin Marunda
Dzidzayi Maruzane
Sipho Moyo
Maxwell Mukwekwerere
Philip Mukwenhu
Tom Muller
Leticia Tapfumanei
Memory Zirobwa

The authors are particularly grateful to the publisher, Ian Murphy, for his enthusiasm and patience in producing this handbook, to Mick Pilcher for his formatting talents and Frances Marks for her editing expertise. The authors wish to thank Brian Williams, Clare Nyandoro and Hannah Jaenicke for their constructive contributions.

The colour photographs used on the cover were taken by and copyight of Richard Barnes.

This publication is an output from a research project funded by the United Kingdom Department for International Development (DFID) for the benefit of developing countries. The views expressed are not necessarily those of DFID. R7275 Forestry Research Programme.

Contents

1 INTRODUCTION

There is very little published information on acacia utilisation and management, and it is perhaps for this reason that they are so widely undervalued.

Trees are an important part of the landscape

This handbook is designed to be used by both agricultural and forestry extension workers, and forestry and conservation teachers in the Southern African region. Although the species and uses are drawn primarily from Zimbabwe, the handbook has application across a much wider area (see inside front cover). It will also be useful to farmers and other people with an interest in growing trees as it contains information on the value, identification, propagation, planting and management of a selection of the most useful indigenous acacia species, plus the closely related *Dichrostachys cinerea* and *Faidherbia albida*. The material provides a basis for a range of training, research and extension activities. It is presented in chapters that can be used either individually or sequentially.

WHY THIS BOOK HAS BEEN WRITTEN

Southern Africa's extensive natural forests and woodlands provide a highly diverse and productive resource; a resource that is being destroyed at an alarming rate. Tree loss is contributing to a range of economic, environmental and social ills. This is felt most directly through widespread shortages of fuelwood, construction material and many other woodland products. Woodland loss has also had a negative effect on the availability of water and soil nutrients on farmlands. Uncontrolled water runoff, following deforestation, has led to widespread erosion and a corresponding shortening of the growing season which has reduced crop yields. Rivers dry out during the dry season because so little water has collected underground. People have to walk such long distances in search of fuelwood that they have little time left to spend on more productive farm activities. These problems are made worse because the firewood shortage has forced people in some places to cook with dung and crop residues, which deprives cropland of a valuable source of nutrients. The rural poor initially feel the effects of deforestation, but the wider effects have serious negative national and regional implications.

Impoverished landscapes follow deforestation

Deforestation has led to land degradation. Both the land and the people are impoverished. There is an urgent need for tree planting. The establishment of trees enriches degraded environments and can increase output from both crops and livestock. Some species that have been selected for this book have attributes that make them particularly suitable for planting on degraded sites. They can help to repair damage to the environment and can lead towards sustainable agricultural production and an improved quality of life.

Enriching the Environment

In the past, farmers were unwittingly discouraged from integrating trees into their agricultural systems. Extension services encouraged, and even insisted on, the total removal of trees from arable lands and the removal of most trees from grazing areas. This resulted in widespread deforestation, particularly in the more populated areas of the country.

The present land degradation in the region is relatively recent when compared to areas like the Sahel. We can learn a lot about our own situation from the experiences in these arid countries to our north. The management of indigenous trees has been critical in the evolution of their agricultural systems. For example, *Faidherbia albida* has been extensively inter-cropped with arable crops in the Sahel. The leaves shed from 50 trees growing in one hectare provide nutrients equivalent to about 50 tonnes of manure a year. *Acacia senegal* forms an important part of the tree fallow systems in Sudan while the gum produced by the trees is especially valuable to the farmers. The long tradition of managing trees in agricultural systems seems to decrease as one comes south through Africa; the value of growing and managing trees has limited recognition in Zimbabwe and adjacent countries. It is common to plant trees around homesteads and in woodlots and orchards. Trees may also be retained around fields. Indigenous fruit trees are seldom cut down and are managed in arable lands. In spite of the use of trees in these situations, the true value of sustaining trees within the arable and grazing areas is seldom recognised and exploited.

Most attempts to counter the effects of deforestation in the past have been based on fast-growing exotic trees such as eucalypts, which are planted in woodlots.

A eucalypt woodlot

This widely sponsored practice has achieved only limited success, mostly because of the costs of establishment, including fencing, chemicals to kill termites and weeding. In addition, eucalypts need a good supply of water and nutrients to grow well. It is now accepted that for people with limited resources, indigenous, pioneer tree species offer an opportunity to counter the effects of deforestation and to increase production from both arable and non-arable lands. The acacias are among the most important pioneer species in the region. Their integration into the farming system is a positive move towards increased and sustainable production.

Why acacias are so important

Acacias are important because they are robust and they provide many benefits. They are strong growers and they are often the only trees found regenerating after deforestation has taken place. Many have the ability to establish quickly and rehabilitate denuded or degraded land. Acacias are generally fast-growing and grow particularly well on bare ground. They are deep-rooted and can reach water and minerals that would otherwise be lost to the system, and release these nutrients to the soil surface through leaf fall. One of their most valuable attributes is their ability to fix nitrogen from the air with the aid of symbiotic bacteria (rhizobia) that live in their roots. These bacteria cause the roots to nodulate and enrich the soil with nitrogen. This is one of the main reasons why some acacias can be grown in close proximity to crops (on contour ridges, along field boundaries and even within the fields) without excessive loss of yield. In addition, some acacias are resistant to drought and termites and their thorns provide a natural defence against browsing livestock. In arid areas, acacias often assume a critical role because they are the main indigenous trees present. They provide a multitude of services and products.

Acacias provide shelter and many useful products

In some areas where they are growing naturally, acacias provide critical life support functions because of their multiple products and services. However, because of the modernisation of agriculture, urbanisation and land pressure, many of the life support functions traditionally provided by acacias are being lost.

THE SPECIES SELECTED FOR INCLUSION IN THE HANDBOOK

Nine species of acacia and two close relatives were selected for inclusion in this handbook, based on their distribution, the importance of their products (firewood, pods, gum, etc.) and how much they contribute to the environment (climate and soil).

The species selected are:

Acacia erioloba
Acacia karroo
Acacia nilotica
Acacia rehmanniana
Acacia tortilis
Faidherbia albida*
Acacia galpinii
Acacia nigrescens
Acacia polyacantha
Acacia sieberiana
Dichrostachys cinerea*

* Dichrostachys cinerea and Faidherbia albida are not acacias but have been included because they share so many characteristics with the acacias and also because of the particular benefits they offer. *Dichrostachys* is distinct because it has different flowers from the acacias. *Faidherbia* was originally grouped under *Acacia* but was changed because of its botanically distinctive characteristics.

When planning tree planting, the easiest way of deciding what trees to plant and which trees require management is to look at what is already growing in the area. If there are people in the area with specialised knowledge of tree planting, they should be approached for advice and assistance.

The usefulness of acacias in farming systems can be increased by:

- learning to identify them and to be aware of their respective uses
- finding out where they grow and how to grow them
- managing both existing and planted trees so as to maximise their productivity

2 ACACIA USES

Acacias provide a wide range of direct and indirect benefits to the household. Products include firewood, fodder, construction material, brushwood fencing, medicines and gum. They also colonise and rehabilitate disturbed sites. They are used to protect gardens and are able to increase agricultural productivity by utilising ground water, cycling nutrients from great soil depth to the surface and by fixing nitrogen from the atmosphere. In addition, they influence the microclimate around them in a way that benefits both crops and livestock. People argue that the biggest problem with acacias is that they are thorny, but it is important to appreciate that the benefits the trees offer far outweigh the disadvantages that their thorns present. The thorns themselves actually protect the plants from browsing animals and dispense with the need for expensive fencing.

The more common uses of the acacias are described in this chapter. These are divided into productive and service uses.

PRODUCTIVE USES OF ACACIAS

FIREWOOD

Most acacias provide good firewood, although their use often depends on what else is available. This is because acacias are thorny and difficult to handle. People get used to the thorns and the wood is easily handled where no other species are available. The best firewood in some areas comes from *Acacia erioloba*, *Acacia nigrescens* and *Acacia nilotica*. These all burn slowly without too much smoke and provide good charcoal. *Acacia karroo* and *Acacia galpinii* also make good firewood and are widely available throughout the country. *Dichrostachys* wood gives a slow-burning fire and can be used to make charcoal. Many people do not like *Acacia erioloba* as firewood because it has an unpleasant smell when it burns, although it makes good coals. *Acacia karroo* burns cleanly with no smell and produces instant heat. The coals do not last long but do make good charcoal. Although it is a fairly light wood, *Acacia polyacantha* makes good firewood and charcoal. *Acacia rehmanniana* and *Faidherbia* do not provide good firewood.

Firewood provides most people in the region with their main energy source

Fodder (browse and pods)

Browse can be pruned to keep it in the reach of livestock

Fodder can be made up of either browse (leaves, shoots, twigs) or pods and is needed throughout the year. Many farmers are unaware of the critical role that acacias play in the natural production of fodder, particularly during that time of the year when little else is available. Acacias are such valuable fodder-producing plants that they deserve careful management.

Browse

Acacia leaves, and the young shoots that have soft thorns, provide a major part of the diet of both domestic livestock and wildlife in the drier parts of the country. The browse available from acacias is particularly valuable when there are no alternatives. This occurs at the end of the dry season and in drought years. Often there is not enough fodder available to carry stock from the end of one season to the time when new growth is available in the next season. This is known as a 'fodder gap'. Acacias such as *Acacia erioloba* and *Acacia karroo* flush before the rains and provide browse during the critical period when there is no grass. Acacia flowers are also an important source of browse, especially the early flowering species such as *Acacia erioloba*. Livestock generally pick them up from the bare ground.

Browse can be lopped off if it is out of the reach of animals

Acacias withstand drought better than grass and often provide the main source of forage for animals during dry periods. During drought periods, branches can be cut so that they fall within the reach of animals.

Thorns protect the acacia from overbrowsing and the grass beneath it from overgrazing

The leaves and twigs of acacias are made up of 7 to 20% protein and a range of trace elements essential for healthy livestock. Secondary compounds such as tannins are present in the leaves and small twigs of some species. These compounds can reduce the digestibility of proteins. There are more tannins in new growth than in mature leaves. This offers the tree some protection from over-browsing during establishment. An even more important defence against browsing livestock is provided by the thorns, which become better developed in the face of heavy browsing. Acacia bushes also give protection to surrounding grasses. The grasses are able to produce seed because of the protection they receive from the overgrowing acacias. This seed is dispersed and helps to re-establish grass cover between the trees when the rains come.

Acacia erioloba, *Acacia nigrescens* and *Acacia tortilis* provide the best browse value of our listed acacias. *Faidherbia* is particularly useful because its leaves grow during the dry season when there is little other browse available. *Acacia galpinii*, *Acacia karroo*, *Acacia nilotica*, *Acacia sieberiana* and *Dichrostachys* are not as widely utilised as the above, while *Acacia polyacantha* and *Acacia rehmanniana* do not have much browse value.

Pods

Acacia pods are a valuable source of livestock feed. Some acacias have seeds which contain up to 28% crude protein. Pods are available from the onset of the dry season. They are easily collected and stored for use as a feed supplement for selected animals, such as milking or in-calf cows, draught oxen, or animals being fattened for sale or slaughter. Ground *Acacia erioloba* pods can also be mixed with maize to make a high quality chicken feed. The pods are produced over an extended season. Some species, especially *Faidherbia*, produce pods right at the end of the dry season when fodder is most scarce. If the rains come early and there is plenty of green grass available, then the animals do not pick up the pods and their nutritional value is lost. In this situation pods can be collected and stored for the following dry season. Goats relish both acacia pods and foliage, while cattle prefer the pods.

Acacia pods and seeds must be crushed if livestock is to be able to benefit from their full nutritional value

In parts of Matabeleland, Zimbabwe, commercial farmers buy acacia pods from communal farmers. This is bad for the communal area livestock because it means they do not have access to the pods, but it does provide the farmers with an immediate cash income. Motivated by the returns from pods, farmers have begun to exercise ownership rights over some pod-bearing trees. Trees have been annexed from grazing land near to people's homes, or farmers have asserted rights over trees in their fields.

Crushed pods provide a valuable feed supplement that can be fed to selected animals

The seed coat of most acacias is very hard. It is necessary to crush or mill the seeds so that livestock can digest them. Seed that is not crushed will be passed out of the animal undigested. The crushed material can be fed by itself, or it can be mixed with crop stover, commercial stockfeed or salt depending on need and availability. Toxins such as prussic acid are sometimes present in the younger pods of *Acacia sieberiana* and *Acacia erioloba*. These toxins can limit livestock intake and reduce protein digestibility. Sulphur can be added to the feed to neutralise the toxins.

Individual large *Acacia erioloba* trees can produce up to 500 kg of pods per year. Acacia pods fall into the High Feed Quality category. When mixed with available roughage they normally fall into the Medium classification.

The most desirable pods for use as feed are those from *Acacia erioloba*, *Acacia nilotica*, *Acacia tortilis*, *Dichrostachys* and *Faidherbia*. Those from *Acacia karroo* and *Acacia sieberiana* have moderate feed value that is marginally better than *Acacia nigrescens* and *Acacia polyacantha*. The pods from *Acacia galpinii* and *Acacia rehmanniana* do not have good feed value.

Feed Quality:

Poor to Medium:	A gut-fill or survival feed during times of drought.
Medium:	Maintenance feed or feed for breeding stock which can be mixed with protein or energy feeds.
Medium to High:	Used as production feed, often when preparing stock for sale or draught use, for lactating cows or weaner feed. This can provide a high proportion of the diet but may require supplements.
High:	Used as feed supplements or concentrates in intensive production such as dairy production and finishing.

NUTRITIONAL VALUE OF ACACIA PODS

Many acacias produce nutritious pods that are an important forage for livestock and wildlife. They are especially valuable because many ripen and fall in the dry season when there is no green grass. At this time they also improve the utilisation of the low quality roughage; all that is available at that time of the year when draught animals must work their hardest.

There is great variation both between species and between individual trees in the time of the year that the pods mature and fall. For example, the pods of *Acacia nilotica* ripen from April to September, those of *A. erioloba* ripen from June to August, those of *A. tortilis* subsp. *spirocarpa* from August to September, and those of *Faidherbia albida* from September to November. Both small-scale and commercial farmers in various parts of Zimbabwe regularly collect pods of these four species and feed them either whole or milled (to release the rich source of protein in the seeds) as a supplement at the end of the dry season. The most useful *Acacia* species in this regard are those with indehiscent pods. These fall to the ground unopened and are much more readily picked up by livestock.

The level of crude protein, one of the main determinants of feed quality, in *A. erioloba* pods is around 10% compared to only 4% crude protein in unmilled maize stover. Maize grain contains about 8% digestible crude protein, whereas milled *A. erioloba* has about 15%, although only 12%of it is digestible. As it is the protein that is costly in a feed, the value of the pods is clear. One drawback is that some unripe pods of this and other species contain concentrations of toxic prussic acid (cyanide), which can affect or kill livestock. Feeding of crushed pods with sulphur locks up any of this compound present into a non-toxic form.

It has been estimated that a parkland with 15 *A. erioloba* trees per hectare can yield an annual pod crop with a crude protein equivalent of 372 kg per hectare, higher than the maximum expected from smallholder grain crops in the same area. There are no input costs associated with this production, except a minimal amount for collection of the pods. Shade, browse and the beneficial effect on the grass production beneath the trees are added benefits.

GUM

Acacia gum is collected from the trunk of the tree

Many acacias produce a gum as a defence against fungal and insect attack. The gum comes out of the tree where the wood is exposed, such as at the site of a wound. The larvae of wood boring beetles tunnelling in the wood readily stimulate gum production in *Acacia karroo*. The gum is edible and possesses both anti-fungal and anti-bacterial properties. The most valuable gum is produced from *Acacia senegal* variety *senegal*, a species which grows naturally from Tanzania northwards. It is not indigenous in South Central Africa, although there are two other varieties of *Acacia senegal* that grow here.

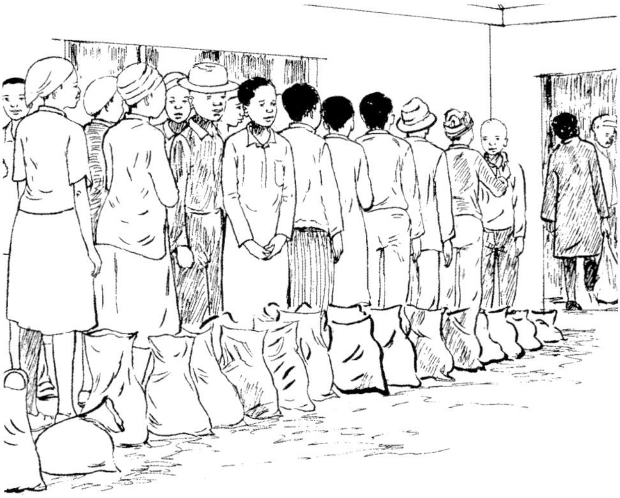

People sell the gum they have collected

Acacia karroo produces a commercially traded gum that is used in the food and pharmaceutical industries. It can also be used in printing inks, for glazing purposes in the pottery industry and in the paint, mining, stockfeed, agrochemical, detergent and glue industries. The gum of *Acacia tortilis* also shows signs of becoming commercially exploitable.

The major constraint in the production of gum in Zimbabwe is the lack of an organised system for production and marketing. At present, people opportunistically collect from the wild and sell to a dealer in Bulawayo. Currently, only one broker is known to be dealing in the gum. The gum is then sold to a few local companies with the surplus being exported to South Africa. *Acacia polyacantha*, *Acacia tortilis* and *Dichrostachys* gums are also collected for chewing. The only country in the region where gum is collected and sold at present is Zimbabwe.

Tannins and dyes

The bark of various acacias can be boiled in water to produce dyes and tannins, which preserve and soften leather. The inner bark of *Acacia karroo* contains up to 19% tannin. It also provides a dye which gives a yellow to brown colour to cloth and a red colour to leather. The bark of *Acacia polyacantha* and *Acacia erioloba* can be used for tanning and dyeing. The bark and the pods of *Acacia nilotica* are used in tanning leather as the pods are rich in tannin and the bark contains 15% tannins. The roots and bark of *Acacia tortilis* provide a yellowy-brown dye.

Acacias provide tannins that are used for curing hides

GUM ARABIC FROM THE ACACIAS

Gum arabic is a substance of ancient international trade in Africa and the Middle East. It has been used traditionally for the mixing of inks and paints, as an adhesive, in cosmetics, perfumery and in medicine, but today it is most widely known as a food additive. Most gum arabic is used to flavour, smoothen, moisten, thicken and stabilize foods and soft drinks. It is also used in the pharmaceutical industry, for example in mixing medicines and making pills.

Gum arabic consists primarily of carbohydrates, (sugars and starches) and their mineral (calcium, magnesium and potassium) salts. It is traded initially as white or yellowish-white spheroidal "tears" of varying size or as angular fragments. It is later sold commercially as white to yellow-white flakes, granules or powder.

Strictly, gum arabic is defined as a dried exudation of gum obtained from the stems and branches of *Acacia senegal* or closely related species. Most of the world supply comes from Sudan; gum arabic accounts for 10-15% of its total foreign currency earnings. Sudan produces gum with the highest and most consistent quality and uniformity because most of it comes from *A. senegal* itself.

Recently, the increase in the consumption of convenience foods has led to an increase in the use of gum arabic and concern about a healthy diet has stimulated an interest in food additives of natural origin. But irregularity of supply and fluctuating prices have caused gum arabic to lose out to man-made substitutes on the world market with the result that annual gum production has dropped from about 70,000 tons in the 1960s to 25,000 tons today. The 1995 price was about US$4200 per ton for ungraded nodules but the prices have fallen over the past three years to about US$2000.

Gums from at least ten species of African acacias have been marketed as gum arabic. *A. karroo* is the best gum producer in Zimbabwe and its gum is widely used regionally as a substitute for gum arabic. But there is some argument over whether this gum is closely enough related to the gum from *A. senegal* and it cannot yet be traded in Europe or the USA. There are many traditional uses of gum arabic in North Africa and the Middle East. It has been an important source of nutrition in the difficult environments of north and west Africa and in the Middle East. For example, in parts of West Africa it is grilled and mixed with butter and sugar to make a sweet which is highly nutritious because of the trace salts it contains. In times of hardship people have been known to live off gum alone for many days. Research has shown that 170 grams is sufficient to support an adult for 24 hours. Gum arabic also has many traditional medicinal uses and, mixed with various substances, it is used to treat headaches, boils and skin rashes, broken bones and breathing complaints. Eating gum is said to strengthen the muscles of the abdomen, to cleanse and purify the digestive system and to serve as a tonic to protect against future disease. Gum arabic has many traditional non-edible uses as well, for example it is used as a starch for the laundering of ceremonial robes.

Gum arabic is a potentially valuable product. With wider acceptance of species source, improved quality control and a more reliable supply, it could provide a cash income for many more resource-poor farmers in Africa.

MEDICINE

Many of the traditional medicinal uses of acacias seem to be based on the astringent properties of the tannins present in the roots and bark. The uses listed below come from a variety of sources and their effectiveness cannot be guaranteed. It must be stressed that extreme care must be taken with all traditional remedies and that they should only be taken under the expert guidance of a traditional herbalist.

Traditional healers use acacia based remedies for their patients

Ailment	**Species**
headaches	burnt *Acacia erioloba* bark *Acacia rehmanniana* roots
toothache	*Dichrostachys* bark
nose-bleeds	*Dichrostachys* roots
discharging and infected ears	crushed *Acacia erioloba* pods
eye, throat and chest pains	*Acacia nilotica* gum *Dichrostachys* roots *Dichrostachys* pods or seeds made into an eyewash
backache	*Acacia rehmanniana* roots *Dichrostachys* roots
rheumatism	*Acacia karroo* roots
general body pains	*Acacia karroo* roots
convulsions	*Acacia karroo* roots burnt roots of *Acacia nigrescens* made into an ointment
open wounds	*Acacia nilotica* roots *Acacia sieberiana*
skin infections	*Acacia tortilis* bark
boils	*Dichrostachys* leaves
scabies	*Dichrostachys* pods or seeds
thrush	*Acacia karroo* gum
swellings	*Dichrostachys* roots and bark
scorpion stings	*Dichrostachys* roots
snakebite	*Dichrostachys* roots *Dichrostachys* leaves chewed and then applied
leprosy	*Dichrostachys* leaves
coughs	boiled *Acacia erioloba* roots mixed with *Opuntia megacantha* root *Acacia nilotica* bark *Dichrostachys* roots

pneumonia	*Acacia karroo* roots *Acacia rehmanniana* roots rubbed into the body *Dichrostachys* roots
asthma	*Acacia karroo* roots and gum
worms	*Acacia tortilis* bark *Dichrostachys* bark
increase urine flow	*Dichrostachys* leaves
stomach pains	*Acacia karroo* roots *Acacia nilotica* gum *Acacia tortilis* bark
upset stomach	*Acacia nilotica* bark mixed with the leaves *Acacia tortilis* bark *Dichrostachys* bark *Dichrostachys* roots
laxative	*Dichrostachys* leaves
infertility	*Dichrostachys* roots
aphrodisiac	chewed *Acacia karroo* roots *Acacia rehmanniana* flowers and leaves mixed
contraceptive	*Acacia karroo* bark fibre tied around a woman's waist *Dichrostachys* roots
depressed fontanelle in infants	*Acacia karroo* gum *Dichrostachys* roots
STDs such as gonorrhoea	*Acacia erioloba* roots mixed with gum *Acacia karroo* bark mixed with ground pods and roots *Acacia nilotica* roots *Dichrostachys* leaves *Dichrostachys* pods or seeds
syphilis	*Acacia karroo* roots

VETERINARY USES	
antidote to 'tulp' poisoning in cattle	*Acacia karroo* bark
infectious cattle diseases	*Acacia karroo* gum burnt in the kraal together with the pods and other herbs
molluscicide to reduce liver-fluke in cattle	*Acacia nilotica* pods in the water control the snails that carry fluke
repel external parasites from poultry	*Acacia karroo* roots spread in fowl runs

Acacia karroo roots spread in a fowl run repel external parasites

TRADITIONAL USES

to ward off snakes in homes and crocodiles at water crossing points	*Acacia polyacantha* roots
children who are restive at night	bathed in *Acacia polyacantha* roots steeped in water
to drive away evil spirits	*Acacia karroo* roots *Acacia nilotica* roots *Acacia rehmanniana* roots mixed with the bark
to protect homes from evil spirits	*Acacia rehmanniana* thorns, leaves and other herbs

BRUSHWOOD FENCING

Acacia brushwood fences are used to keep livestock out of cropped areas

Acacia branches are often used for brushwood fencing around gardens, homesteads and fields as they can be very effective in excluding livestock. Cut branches are stacked against each other around the area that is to be protected. This is easier and more effective when stacked along a barbed wire fence. In places where predators like lions are a problem, acacia branches are piled against the cattle pens to provide protection. Brushwood fencing is generally used for firewood the following dry season before it has time to rot. Thorn branches can also be used to protect individual seedlings.

Acacia karroo, *Acacia tortilis* and *Dichrostachys* have effective thorns and are the most popular for brushwood fencing. *Acacia nilotica* and *Acacia rehmanniana* are also useful because of their strong thorns. *Acacia tortilis* and *Acacia karroo* are rapidly attacked by borers and fungi and they have to be replaced more often than *Dichrostachys*, which has very hard and durable thorns.

Some acacias provide useful fence poles

FENCING POLES

The best fence posts are durable and resistant to borers and termites. These include *Acacia nilotica*, *Acacia erioloba*, *Acacia galpinii*, and *Acacia tortilis* and *Acacia nigrescens*, of which the latter is the most valued, particularly for hut and kraal poles and for fence posts. The bark is removed from fencing poles to reduce borer attacks. The proportion of hardwood to sapwood is important becasue the sapwood is not durable. Although some of the other acacias also produce straight stems, they are not durable. Charring the ends of the poles that are to be set into the ground provides them with additional protection from termites. *Dichrostachys* has a dark brown wood that is strong and hard that is used for small fence posts or droppers.

Acacia nilotica seasons well, is durable and is resistant to borers and termites. It is commonly used for fence posts and mine timber in spite of its crooked form.

Timber or rough construction wood

Acacia timber can bear heavy loads

Acacia timber is widely used in areas that acacias have colonised and where they are the dominant species. The density and weight, which is linked to durability and size, determines suitable uses. Some species, such as *Acacia nigrescens* and *Acacia erioloba,* have very heavy wood that is both durable and suitable for outside construction as well as providing excellent planks, although they are very hard on saw blades. *Acacia nigrescens* is a widely distributed species that grows comparatively straight and has a reasonable quantity of heartwood when pole-sized. The heavy, dark brown heartwood is very strong and durable, but tends to crack and is difficult to work. It is resistant to borer, termite and fungal attack. *Acacia galpinii*, although fast growing, provides quality planks. Owing to its size, it gives possibly the highest timber yield of all Zimbabwe's acacias (up to 2.5 cubic metres from large trees). The hard, heavy heartwood is durable, but it is not suitable for outdoor use. *Acacia polyacantha* provides good timber but needs to be seasoned slowly to avoid splitting.

Seasoning involves either immersing the timber in water for periods of no less than two weeks, or by slow air drying under shade. Seasoning in water produces a timber that has better resistance to borers and termites, while air-drying makes the timber less likely to split and twist after cutting. Debarking the trees is required for water treatment. Another way to reduce insect attack is to burn off the bark.

Some of the acacias are unsuitable for timber either because the wood is of small size, because it is mainly sapwood, or because it is liable to split and twist.

Furniture timber

Some acacias provide useful timber for furniture making

Acacia galpinii provides a useful timber for furniture production. *Acacia karroo* is also used, although the timber is generally unsuitable for outside use in particular as borers and fungi rapidly attack it. Though it is a tough and resilient timber, it is liable to spit and twist. Furniture is also made from *Acacia polyacantha*, which has a red-brown heartwood that is moderately heavy and durable. It is prone to splitting and needs to be seasoned slowly to prevent this. After seasoning, *Faidherbia* timber is valued for making doors and tool handles. Large trees are also made into dugout canoes.

Carving and turning wood

Acacia nigrescens, which is resistant to borer, termite and fungal attack, is often carved. *Acacia nilotica* is very durable and resistant to borers and termites and is used for tool handles. *Acacia erioloba* is also used for carving. Though small in size, *Dichrostachys* wood is very tough and is used to make tool handles and assorted household utensils.

Acacia wood is used for tool handles or carving

Bark fibre

The inner bark of acacias provides a strong fibre

The inner bark of *Acacia karroo*, *Acacia erioloba*, *Acacia nigrescens*, *Acacia polyacantha*, *Acacia rehmanniana*, *Acacia tortilis* and *Dichrostachys* are all used for string. *Acacia karroo* is the most preferred of these and is also used for basket-making. Stems that have had their bark harvested should be cut from the plant. This encourages vigorous regrowth and minimises the possibility of the plant becoming infected through large open wounds.

Bee forage

The flowers of most acacia species, especially *Acacia karroo*, contain good quantities of nectar. There is the potential to have many more beehives where there are a variety of acacias as they provide both quality and quantity of pollen and nectar. Acacia honey tends to be runnier than that of other species and it has a strong but not unpleasant taste. *Acacia karroo* and *Acacia tortilis* are particularly important for honey production due to their conspicuous flowers. *Acacia galpinii*, *Acacia nigrescens*, *Acacia nilotica*,*Dichrostachys* and *Faidherbia* all produce good quantities of both nectar and pollen. *Acacia galpinii* and *Acacia nigrescens* flower from August to September, *Acacia nilotica* from October to January, *Acacia karoo* from November to January, *Acacia tortilis* from November to March, *Dichrostachys* from November to January and *Faidherbia* from May to July. There is no record of bees being seen on the flowers of *Acacia erioloba*.

Edible seed

Faidherbia seed is consumed in areas around Binga when there is an acute food shortage. Many households collect and store seed for future consumption. The preparation is laborious and can take up to two days. The seed has to be boiled several times. This boiling allows the removal of the skins. To reduce the bitterness of the seed it is then boiled with ash and the water is again discarded.

The seeds of *Acacia erioloba* and *Acacia karroo* are roasted and ground to make a coffee substitute.

Service uses of acacias

Live fences and hedges

Live fences or hedges provide an alternative to boundary fencing as they can be used to protect areas from the damaging effects of roaming livestock. However, they generally require some years to grow into an effective barrier. Trees can be grown so that they form either live poles to carry wire or horizontal poles or hedges which form barriers to livestock.

A pruned acacia hedge can keep out livestock without using wire

Acacia tortilis, *Acacia karroo* and *Dichrostachys* make good live fencing around gardens, homesteads and fields because of their form and thorniness. *Acacia nigrescens*, *Acacia nilotica* and *Faidherbia* can also be used as live fence poles. Other species that can be combined to increase the efficiency of living fences include *Agave americana*, *Euphorbia tirucalli*, *Kirkia acuminata*, *Jatropha curcas*, and a variety of *Ficus* and *Commiphora* species.

Boundary markers

Planted acacias can stand out in such a way as to demarcate fields, paddocks or other boundaries. This demarcation can be to minimise disputes over field boundaries between neighbours, or to demarcate paddocks to allow for herded rotational grazing. Some of the acacias are additionally useful if the demarcated area is near fields because of the positive effects they have on crops.

Pollarded acacias provide live fence poles and a source of fodder

Shade and ornamental

Most acacia trees provide shade, the value of which is often underestimated. Animals use energy to keep cool, and if they can rest in the shade in the heat of the day, that energy is put into weight gain. Trees also reduce loss of heat by radiation from the earth on cold nights and so keep the animals warmer. Then the animals do not have to use as much energy in keeping warm. Shade at water points reduces livestock water requirements and intake. Shaded pens provide protection from the sun, rain, wind and hail. It has been shown that protection provided by trees can increase the conception rate of cows.

Acacias are very useful for shade and protection

Ornamental and shade trees provide useful by-products such as poles, fodder and fuelwood. In denuded areas, they also provide some protection from lightning if they are higher than the surrounding homes.

Acacia erioloba, *Acacia galpinii*, *Acacia karroo*, *Acacia nilotica*, *Acacia sieberiana*, *Acacia tortilis* and *Faidherbia* are all suitable shade species. *Faidherbia* is sometimes planted as an ornamental as it is in leaf during the dry season and *Acacia xanthophloea* (although not one of our selected species) is widely planted as an ornamental because of its unusual yellow-coloured bark.

Soil fertility and enrichment planting

Acacia karroo and *Faidherbia* can both improve soil fertility and facilitate the establishment of nutritious grass species under their canopies. Acacias have the ability to utilise water and nutrients from deep in the soil. They fix nitrogen, improve soil structure and aid the infiltration of rainfall. Their leaf litter also improves the soil. The soil enrichment capability of *Faidherbia* is very important in areas where the soils are infertile and where soil moisture is available deep down in the dry season. The species sheds its leaves at the start of the rainy season. This reduces competition with crops for soil moisture and light. Research in West Africa has shown that *Faidherbia* can increase grain yields by at least 40% in comparison to yields in fields without *Faidherbia* and where no fertiliser was applied.

Grass under acacias is often more nutritious than grass growing in the open

Palatable grass species, including *Panicum maximum*, flourish under many of the acacias. This is because of the nutrient input from the acacia leaf fall and the partial shade that they provide. Humus is sometimes collected from under acacias such as *Acacia karroo* and transferred to gardens to improve the soil.

Tree fallow cropping

Fallow land may be planted with one of the selected species to hasten the rehabilitation of areas where fertility has dropped to unproductive levels for cropping. This is of particular benefit to poor farmers who have fields that have been forced out of production because of a decline in soil fertility. During this period, the trees also enhance the value of the area by producing livestock browse and encouraging more nutritious grasses. The trees may be in pure or mixed stands.

Species selected should be fast growing, easy to establish and to remove and should be nitrogen-fixing. Some of the acacias, especially Acacia karroo, Acacia polyacantha and Acacia tortilis have these attributes and can be used.

Soil conservation

Acacias can be used to rehabilitate gullies

Acacias can play an important role in the reduction of soil erosion. In many harsh environments the ability to establish readily and to persist and spread make acacias important species for rehabilitating eroded sites. The beneficial effect of species like Acacia karroo on grass growth and its soil-holding properties can be exploited. Management practices such as thinning and pruning improve tree-grass interaction, hasten the rehabilitation of degraded sites and also improve grazing at the same time. Acacia karroo and Acacia tortilis are being widely used in the rehabilitation of mine dumps due to their robustness. Acacia polyacantha, Acacia nilotica, Acacia rehmanniana, Acacia sieberiana, Acacia tortilis, Dichrostachys and Faidherbia are also planted for soil conservation purposes.

Windbreaks

Windbreaks can provide protection to homesteads, crops and livestock

The damaging effect of wind, through the removal of the fertile top layer of soil, is often cause for concern. The use of trees planted across the direction of the prevailing wind can reduce this problem. Many trees can be used for this but Acacia polyacantha and Acacia karroo are particularly useful in the creation of effective windbreaks. Their beneficial effects on neighbouring crops make them suitable species to plant as windbreaks near fields.

3 Identifying Planting Sites

Give consideration to all planting sites

Acacias are very useful trees. They can be grown in a wide range of situations. Even in the driest or the most intensively cultivated and heavily populated areas there are places where acacias can be grown without interfering with cropping or other on-farm activities. This chapter considers tree-planting opportunities that exist in commonly available planting sites.

A farm survey on page 93 identifies many planting sites where trees can be grown. The tree data sheet on page 94 will help to identify them and include the following:

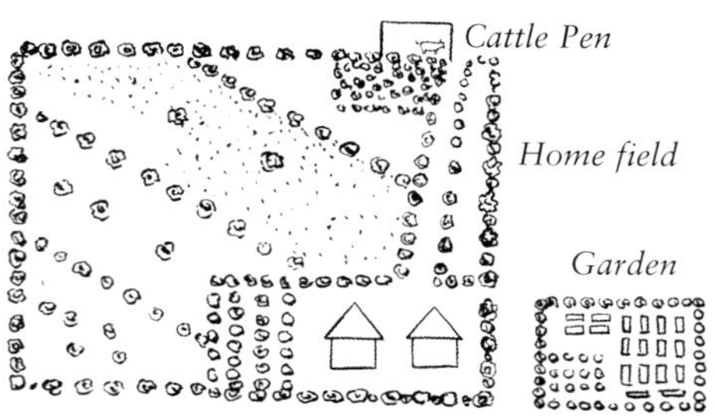

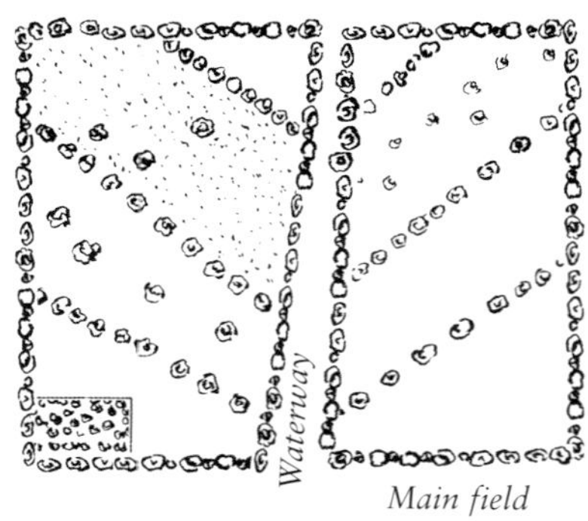

Homestead

Trees provide shade at the homestead

This is the cleared or swept area around a house. Acacias can be planted in this area for shade or ornament. They are often removed before their thorns grow out of reach.

They are sometimes grown around the homestead as live fences to help keep out animals.

Arable fields

These usually include a homefield, which is a cultivated or fenced area around the house, and a mainfield, which is usually some distance from the house and is less likely to be fenced. Planting sites in fields can be divided into:

Contours

Contours account for up to 10% of a field's area and provide significant opportunities to plant acacias along the ridges. *Acacia galpinii* and *Faidherbia albida* can be considered, with other fruit and fodder species interplanted between them.

Trees can make a contour ridge productive

Withinfield

Planting in fields can be done as an intercrop or in a block. Intercropping can be trees scattered or planted in lines within arable lands. Blocks are usually established in corners of the field that are difficult to crop.

Along paths, roads or field boundaries

The sides of all paths and roads provide opportunities to establish avenues. Trees can also be planted along field edges where they can define the field's boundary.

Some acacias improve soil fertility

Trees can protect paths and fields

In stony or waterlogged areas

In these areas, crop yields are often too low to be economically viable. With careful selection of species, these areas can become important to the household economy.

Trees can make stony or waterlogged areas productive

Worn-out fields

Fields not cultivated because they have lost fertility can be rehabilitated by using an acacia tree fallow.

Gardens

These are generally well protected areas with access to water all year round.

Non-arable areas

Non-arable or grazing lands are severely degraded in some highly populated parts of Southern Africa. Tree cover is often reduced from the original 70 to 95% down to less than 5%. Woodland in these areas normally varies from poor scrub to well-established mature, scattered trees. However, loss of tree cover, together with the damage caused by over-grazing and subsequent erosion, has reduced the productivity of these lands to well below their site potential. Apart from this, tree cover facilitates water infiltration. The extensive removal of trees has had a severe negative effect on the water cycle in these areas. In order to improve both the productivity and the hydrology of degraded non-arable lands, the extensive re-planting of trees is an urgent environmental priority.

Gardens provide planting and nursery sites

The aim should be to establish an artificial savanna woodland with a tree canopy cover of between 30 and 50%. This mimics the ecology of the natural savanna woodland that once existed on the site.

Trees play numerous roles in non-arable areas

The species selected should be fast-growing, provide multiple products and services, and be easily established with minimum inputs. The 11 acacia species discussed have these characteristics and offer many opportunities for planting in non-arable lands.

Acacias are useful to help reclaim degraded land

Degraded non-arable areas include grazing lands, hills, rocky outcrops, river valleys and vlei margins. They offer many opportunities for planting and managing acacias. The communal management generally applied in these areas suggests that the activity needs to involve many people. Community leaders should be involved as early as possible in the exercise. The process should involve discussion with the community in order to ensure active and continued group participation. Non-arable areas provide an opportunity for concerted community-based actions. These will rely on a dynamic community spirit or on state-sponsored activities such as food-for-work programmes. These activities can make significant improvement to grazing areas through the management and use of existing acacias and by planting suitable additional ones.

River valleys

In Zimbabwe, the natural vegetation along watercourses is protected by law. In practice, the law is often not enforced and much of the riparian vegetation in highly populated areas is extensively degraded through excessive erosion and the collapse of stream and riverbanks. Seven of the species selected for this book occur naturally in riverine fringes and are useful for stabilising degraded riparian environments. These species include *Acacia galpinii*, *Acacia karroo*, *Acacia tortilis* and *Faidherbia albida*.

Other planting sites

Trees help stabilise river banks

The list of planting sites can be added to in consultation with the community. Other planting opportunities include gullies, mine dumps, bus stops, market areas, water points, church lands, schools and colleges.

When acacias are being planted to provide a particular service rather

Acacia trees and grasses are used to stabilise minedumps

than a product, the planting site is dictated by the service that is expected. For example, a live fence or windbreak will be planted along a boundary, shade trees will be planted where the shade is required, and trees that improve soil fertility will be planted in the area where the soil is to be improved.

Nature reserves

Areas of relatively undisturbed indigenous vegetation are extremely rare in highly populated areas. Where they still exist, they are normally confined to rocky hills and outcrops (kopjes) or otherwise remote, inaccessible sites. Where such vegetation is found, be it woodland, wetland or riparian forest, efforts should be made to protect it as a Nature or Biodiversity Reserve. These areas will require designation, protection and maintenance to ensure the preservation of seed supplies of trees best adapted to the area. These resources are an irreplaceable national heritage, in addition to being aesthetically pleasing. Traditionally, sacred hills, forests and woodlands as well as graveyards provide possible sites for this purpose. Local spirit mediums are often already or willing to become custodians of such places.

Nature reserves are an invaluable gift to the future

Woodlot planting

Acacias are seldom established in woodlots or plantations in Zimbabwe. However, acacia plantations may become more common in future. One potential acacia plantation species is *Acacia senegal* variety *senegal* from North Africa, which is being tested in Zimbabwe. The trees improve soils, are browsed and produce a valuable gum. As the demand for poles, firewood and other tree products increases, some of our local acacias may also become commercially viable in plantations. The growth rate of some acacias, such as Acacia karroo, suggests that they are more productive in plantations than has previously been thought.

Species and provenance trials of African acacias

Man's progress is based upon increased production and quality of food, fibre and fuel crops. Genetic improvements have been achieved through selection and plant breeding. Records show that the main crop species represent vast improvements over their wild ancestors. There is also still enough genetic variation left in these species for selection and breeding to make even greater contributions to productivity in the future.There has been little selection and breeding of tree species and few are more than one or two generations removed from the wild undomesticated state. The potential for improving tree characteristics has already been shown in breeding programmes for the pines and eucalypts where the value of the crop can be more than doubled in just a single generation of selection.

The African acacias are important contributors to agricultural production even in their wild state. It is known that they are extremely variable, in growth rate and pod production. They are, therefore, ideal subjects for rapid genetic improvement through modern methods of selection and breeding. This has been recognized by international and national agencies and projects have been started to test and use this variation to improve the productivity of some of the most important acacias, for example, *Acacia erioloba*, *A. karroo*, *A. nilotica*, *A. senegal*, *A. tortilis* and *Faidherbia albida*. All these species occur naturally in Zimbabwe.

These projects have been coordinated by the Oxford Forestry Institute. Seeds from 141 different localities (provenances) of the six acacia species have been collected across their whole range in Africa. The Zimbabwe Forestry Commission have provided a centre for the first field trials which have been established over 14 different sites in Matabeleland where the acacias are important to small-scale farmers. Results of these trials show wide genetic variation in these species. Large differences in survival, growth rate, pod production etc. between species and provenances on the same site have been shown.

Among the best overall performers so far under these conditions have been the Chivu (Zimbabwe) provenance of *A. nilotica* subsp. *kraussiana*, the Maun (Botswana) *A. tortilis* subsp. *heteracantha*, the Devure (Zimbabwe) *A. tortilis* subsp. *spirocarpa* and the Jedibe Island (Okavango, Botswana) *A. karroo*. The fastest-growing *A. erioloba* comes from Zambia and the worst from the northern Cape Province of South Africa. The high altitude provenances of *F. albida* from Ethiopia have done best on the cold sites whereas some of the local provenances of this species, such as Gonarezhou (Zimbabwe) and Kapula (Zimbabwe), have done best where there is no frost.

Significant information is coming from these trials and over the next ten years it will be possible to make recommendations on which provenance of which species should be planted in which agricultural niche and in which environment to provide a particular product or service most efficiently.

4 WHICH ACACIAS TO PLANT

Once you have decided to plant acacias, identify the best species to use. The choice will depend on why the trees are to be grown, where they are to be grown and what is available. Always look at what is already growing before making your choice.

This chapter outlines a process to choose species for a particular planting area. The exercise can lead to a list of species that should produce what is wanted on the available planting area. The process involves **six steps**. An example of the procedure is provided on page 38.

Wherever possible, plant species that grow naturally in a particular ecozone, although there can be exceptions to this rule. In nature, plants compete with each other to grow on a specific site. A species occurs naturally in an area where it competes well enough to establish itself over successive generations. The level of competition with other plants is often most critical during establishment. If human assistance is given, a species may be able to thrive outside the ecozone in which it is normally found. This means that if one of the species covered in this handbook produces a desirable product it might be possible to grow it outside its usual range. It is desirable, however, to use seed collected from an ecozone with as close a match as possible to the proposed planting area. Do not try to establish species in areas that are very different from where they normally grow. For example, it is not recommended to try to establish *Acacia polyacantha*, *Acacia rehmanniana* or *Acacia sieberiana* in lowveld areas with low rainfall, or to try to grow *Acacia erioloba* or *Acacia tortilis* in the moist highveld, or to grow *Faidherbia albida* above 1200 metres, where it produces very few pods.

STEP 1 Identify your ecozone and the species that grow there naturally

The area covered by this handbook can be divided into five ecozones. These, and the selected acacia species that grow in each ecozone, are shown in **Table 1**.

Look at **Table 1** and identify the ecozone where the trees are to be planted. You can also look at the distribution maps for the individual species and the list of acacias from different habitats on page 134. *In Row 1 of* Table 6 *place a tick in the column for each species that grows in the ecozone of the selected planting area.*

STEP 2 Identify what uses you require from the acacias

The main reason that people grow plants is for the value of the products or other benefits that they provide. **Table 2** details uses of the selected acacias.

In Table 2 *there is a column marked* 'Priority'. *In this column, mark with a pencil which of the listed products or services you require.* The number of uses chosen is up to the user and can be from just one use to all of them.

For each species that you entered in Row 1 of **Table 6** look down the columns of corresponding species in **Table 2** and add up all the figures (1 to 3) for each of the uses you ticked in the column marked **'Priority'**. *Write the totals in the bottom row of* Table 2. These totals indicate which species best produce what is wanted. *Enter these totals in Row 2 of* Table 6. You might at this stage want to exclude any species that are not widely recognised for specific uses. This is easily done by ignoring all species that score 2 or less when transfering the totals to **Table 6**.

STEP 3 Select species suitable for the soils at the planting area

Eliminate any species that do not grow well under the soil conditions that occur in the planting area. We can do this by looking through **Table 3**. *Tick the soil characteristics of the planting area in the appropriate column in* Table 3. (It might be necessary to tick more than one line. For example, there might be heavy soil subject to seasonal waterlogging in which case you would tick the rows for heavy soils and for seasonal waterlogging). *Then look at each species listed in Row 1 of* Table 6 *and place a tick in Row 3 of* Table 6 *for any species shown in* Table 3 *to occur naturally in these soil types.*

STEP 4 Select the growth features you require for species in your planting area

Look at Table 4 *and put a tick in the priority column for each of the growth features you require.* Only tick the column for frost, termite or drought resistance if it is a known problem in your area. *Give preference to species that coppice or grow fast. In* Table 4, *for each species you have selected for your planting site (those you have entered in Row 1 of* Table 6*) add up the total (numbers 1-3) that occur in the rows which you have selected and enter the totals in the last row of the table. Enter these totals in Row 3 of* Table 6. You might at this stage want to specify that you require a grade 3 rating for a particular feature, for example termite resistance. If so, ignore all species that score 2 or less for this attribute and omit the figures for these species when you transfer the totals to **Table 6**.

STEP 5 Select the site for your planting

Identify the site of your planting area and tick the relevant line in the priority column of Table 5 *to describe it. Look down the columns of* Table 5 *at the species you have chosen in Row 1 of* Table 6 *and put a tick in Row 5 of* Table 6 *for each species that is shown as suitable for the site of your planting area.*

If you want to plant in more than one site in your planting area, you should perform the whole selection procedure again for the second site because the same species may not be suitable for both sites.

STEP 6 Identify suitable species to plant

Now that the first five rows of Table 6 *have been completed, total the numbers for each species and write them in Row* 6. Do not write in totals for any of the species that do not have a number or a tick in all Rows from 1 to 5. Those species that have a total in Row 6 of **Table 6** are possible choices for your planting area. They can be ranked for suitability according to the totals in Row 6.

Remember that wherever possible you should discuss your plans with anybody from the community who has local knowledge and relevant expertise or experience with acacias.

Table 1. Ecozones where the selected acacias occur naturally

Ecozone	Highveld	Middleveld (High rain)	Middleveld (Low rain)	Lowveld (High rain)	Lowveld (Low rain)
Altitude	over 1200m	750-1200m	750-1200m	below 750m	below 750m
Rainfall	over 800mm	over 650mm	below 650mm	over 650mm	below 650mm
Acacia erioloba		🌳	🌳		
Acacia galpinii	🌳	🌳	🌳	🌳	🌳
Acacia karroo	🌳	🌳	🌳		
Acacia nigrescens		🌳	🌳	🌳	🌳
Acacia nilotica	🌳	🌳	🌳	🌳	🌳
Acacia polyacantha	🌳	🌳	🌳	🌳	
Acacia rehmanniana	🌳	🌳	🌳		
Acacia sieberiana	🌳	🌳	🌳	🌳	
Acacia tortilis		🌳	🌳	🌳	🌳
Dichrostachys cinerea	🌳	🌳	🌳	🌳	🌳
Faidherbia albida		🌳	🌳	🌳	🌳 *if ground water available*

Key: 🌳 Areas where the species are naturally found

Table 2. Uses of acacias
Production and service

Use	Priority	*Acacia erioloba*	*Acacia galpinii*	*Acacia karroo*	*Acacia nigrescens*	*Acacia nilotica*
Firewood		3	1	2	3	3
Charcoal		2	1	3	2	2
Fodder – leaves		2	1	2	2	1
Fodder – pods		3	1	2		3
Brush fencing		1		2	1	
Fencing poles		3	1		3	2
Construction		2	2	1	3	1
Furniture			2	1		
Carving wood			1		1	
Gum		1		2		1
Edible seeds		1		1		1
Medicine		1		2		1
Bee forage			2	3	1	1
Tanning				1	1	2
Dyes				1		1
Bark fibre		1		2	1	
Live fencing				2	2	1
Shade		3		1		2
Ornamental		2	2	2		
Soil fertility improvement		1		3	1	1
Soil conservation				2	1	
TOTALS						

Key: 3 = very suitable 2 = suitable 1 = marginally suitable

Table 3. Soil types on which acacias are normally found

Soil type	Planting site	*Acacia erioloba*	*Acacia galpinii*	*Acacia karroo*	*Acacia nigrescens*	*Acacia nilotica*
Light (sandy)		●●				●
Medium (loamy)			●●	●●	●●	●●
Heavy (clayey)			●	●	●	●
Shallow/stoney				●	●	●
Saline				●		●
Seasonally waterlogged						●
Wet area margins			●	●●		●
Riverine			●●	●●	●●	
Termitaria					●	●●

Key: ● = suitable ●● = very suitable

Acacia polyacantha	*Acacia rehmanniana*	*Acacia sieberiana*	*Acacia tortilis*	*Dichrostachys cinerea*	*Faidherbia albida*
1		1	2	3	
			1	3	
			2	1	3
		1	3	3	3
	1		3	1	
			1	1	
2				1	
2		1			1
1			1	1	
1				1	
					1
1	1	1		2	1
1			2		3
1				1	
1			1		
1	1	1	2	1	
			2	1	
		3	1	1	2
1		2			1
		2	1	1	3
2	1	1	1	1	3

=Service uses

Acacia polyacantha	*Acacia rehmanniana*	*Acacia sieberiana*	*Acacia tortilis*	*Dichrostachys cinerea*	*Faidherbia albida*
		●	●●	●●	●
●	●	●●		●	
●●	●		●	●	
	●●			●●	
				●	●
●●	●●	●●	●		●
●●	●●	●●	●●		●
●●		●	●●	●	●●
				●	

Table 4. Growth features

Features	Priority	*Acacia erioloba*	*Acacia galpinii*	*Acacia karroo*	*Acacia nigrescens*	*Acacia nilotica*
Frost tolerance		3	2	2	1	1
Termite resistance		3			2	1
Drought tolerance		3	2	3	3	3
Coppice ability		1	1	2	1	1
Growth rate		1	2	3	2	2
Total						

Key: 3 = very suitable 2 = suitable 1 = marginally suitable

Table 5. Suitable planting sites

Site/Species	Priority	*Acacia erioloba*	*Acacia galpinii*	*Acacia karroo*	*Acacia nigrescens*	*Acacia nilotica*
Homesteads		●	●			●
Contours		●		●		●
Within field						
Along paths, roads or field boundaries		●	●	●		●
Non-arable areas in fields				●	●	●
Worn out fields – tree fallows				●		
Gardens						
Non-arable areas or grazing land		●		●	●	●
Eroded areas/gullies				●		
River valleys			●	●	●	●
Sodic or otherwise toxic soils				●		●
Waterlogged areas/ vlei margins				●		
Termitaria					●	

Table 6. Selection and elimination of species

	Acacia erioloba	*Acacia galpinii*	*Acacia karroo*	*Acacia nigrescens*	*Acacia nilotica*
1. Ecozone (Table 1)					
2. Uses(Table 2)					
3. Soil (Table 3)					
4. Growth (Table 4)					
5. Site (Table 5)					
6. TOTALS					

Acacia polyacantha	*Acacia rehmanniana*	*Acacia sieberiana*	*Acacia tortilis*	*Dichrostachys cinerea*	*Faidherbia albida*
1	1	2	1	1	
		2	2		
2	2	3	1	1	3
1	2	2	1	3	2
3	2	2	1	2	3

Acacia polyacantha	*Acacia rehmanniana*	*Acacia sieberiana*	*Acacia tortilis*	*Dichrostachys cinerea*	*Faidherbia albida*
		●			
					●
					●
●		●	●	●	●
●			●		●
●			●	●	●
					●
	●		●		●
●			●	●	●
●		●	●		●
			●	●	
●	●	●			●
	●				●

Acacia polyacantha	*Acacia rehmanniana*	*Acacia sieberiana*	*Acacia tortilis*	*Dichrostachys cinerea*	*Faidherbia albida*

EXAMPLE OF THE SPECIES SELECTION PROCEDURE

This will indicate what acacias should be planted in a high rainfall middleveld area to produce firewood and fodder (leaves and/or pods) and to enhance soil fertility on a heavy soil where there is frost in most winters and the site available is non-arable grazing land.

STEP 1. Identify your ecozone and the species that grow there naturally
In Table 1 all species are shown as suitable for the high rainfall middleveld. So tick all species in Row 1 of Table 6 as shown in the replica of Table 6 below.

STEP 2. Identify what products or services you require from the acacias
In Table 2 place ticks in the priority column for the rows for firewood, fodder - leaves, fodder – pods and soil fertility improvement. Then add up the numbers in the columns for each species that produces firewood and fodder, and improves soil fertility improvement and enter the totals in the last row of Table 2. Transfer the results to Table 6, as shown below.

STEP 3. List the species that are suitable for the soil type
In Table 3 place a ✔ by heavy soil. Tick the columns in Row 3 of Table 6 for each species that grow in heavy soils as shown below.

STEP 4. Select the growth features you require
In Table 4 place ticks in the priority column in the rows for frost tolerance, and fast growth. Then enter the total in the bottom row of Table 3 for each species that is both frost tolerant and fast growing. Transfer the results to Row 4 of Table 6 as shown below.

STEP 5. Select the site for your planting
In Table 5 place a tick in the priority column in the row for non-arable areas or grazing land and tick in Row 5 of Table 6 for each species where a dot occurs in Table 5 as shown below.

STEP 6. Identify suitable species to plant
Now you have filled in rows 1 to 5 in Table 6, enter the total (Row 6) just for those species that have either a number or a tick in all five rows. Acacia karroo, Acacia nigrescens and Acacia nilotica, are the three species that fulfil all your requirements for your planting area. Acacia karroo looks to be the most suitable because it has the highest total (14) in Row 6. By studying the blank cells in the table and the relative size of the numbers, you can review why the species were eliminated or were deemed less suitable than Acacia karroo.

EXAMPLE OF COMPLETED TABLE 6. SELECTION OF SPECIES

	Acacia erioloba	Acacia galpinii	Acacia karroo	Acacia nigrescens	Acacia nilotica	Acacia polyacantha	Acacia rehmanniana	Acacia sieberiana	Acacia tortilis	Dichrostachys cinerea	Faidherbia albida
1. Ecozone (Table 1)	✔	✔	✔	✔	✔	✔	✔	✔	✔	✔	✔
2. Uses(Table 2)	9	3	9	6	8	1		4	8	7	9
3. Soil (Table 3)		✔	✔	✔	✔	✔	✔		✔	✔	
4. Growth (Table 4)	4	4	5	3	2	3	3	5	2	2	
5. Site (Table 5)	✔		✔	✔	✔		✔				✔
6. TOTALS			14	9	10						

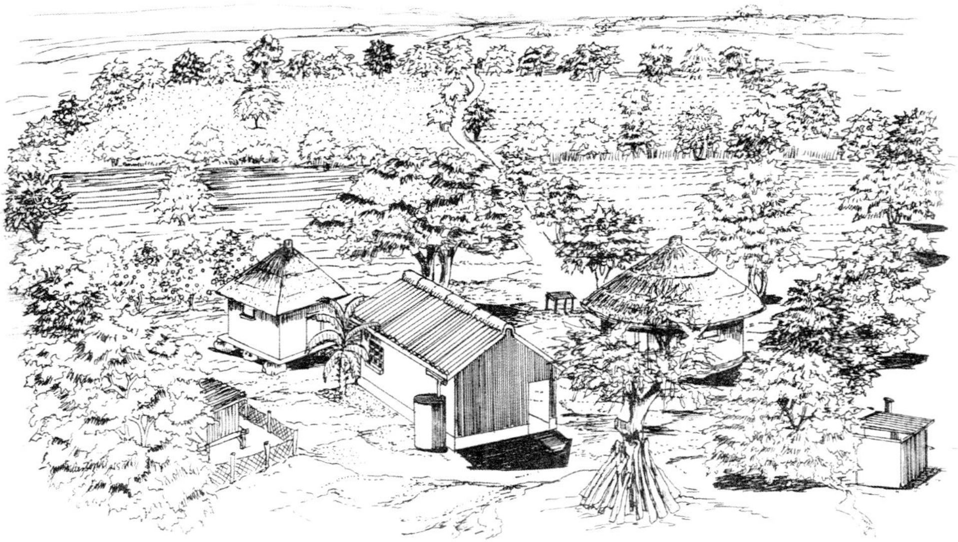

When choosing where to plant acacias this guide will help to identify available sites.

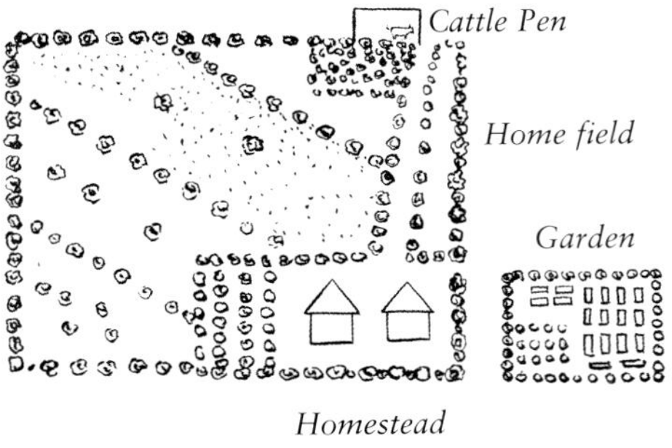

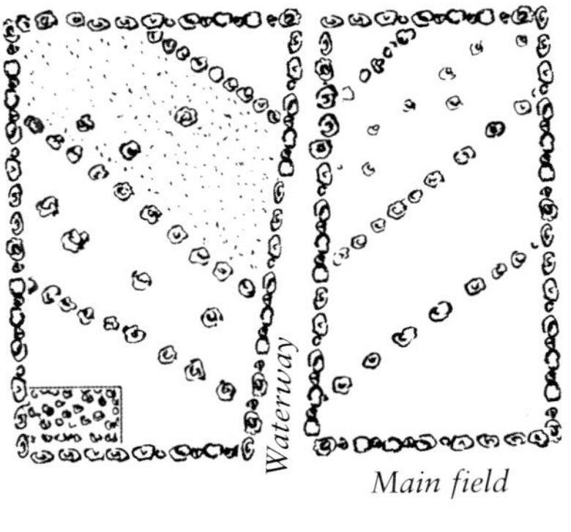

5 Raising acacia seedlings

Seedlings need a regular supply of water and protection from livestock

The high cost of buying tree seedlings makes growing them yourself an attractive alternative. This is only possible with a regular supply of water and an area that is protected from livestock. This chapter offers guidelines on seedling production.

To grow seedlings effectively, it is necessary to start nursery preparation at the end of the previous rains.

Planning

Having decided which seedlings to plant, it is necessary to establish how many seedlings are required. Start with a visit to the planting site. Measure the area available to plant and match this with the layouts suggested on page 65. Calculate how many seedlings are needed in each of the sites that are to be planted. Decide how much of the total will be planted each year. Do not try to do too much at once, spread the load. Seed is best sown after the last frost of the season.

Seed

Sources of acacia seed

The cheapest seed is collected directly from the trees.

Seed is often available from National Seed Centres (page 57). One advantage of getting seed from a supplier is that they may be able to provide improved seed that has either a high growth rate or some other desirable features.

Seed can be purchased from a supplier

Collecting acacia seeds

When collecting seed it is important to collect from healthy trees. When selecting the trees, choose those with the shape or form of the trees that are wanted. As an example, if the trees are to be grown for poles then collect the seed from trees that have particularly straight stems. If pod production is the purpose of growing the trees, then collect seed from trees that produce many pods. Avoid collecting seed from isolated, poorly shaped, diseased, old, or low seed-producing trees. The best trees for seed collection are those growing in the midst of a healthy stand of the same species. Try to collect seed from at least 20 trees and mix it in roughly equal quantities in order to preserve genetic variation for selection and conservation.

Collect only mature seed. Seed collection from dehiscent (splitting) pods must be carefully timed. Avoid collecting unripe seed or pods that have already shed their seed. Cloth bags that allow free exchange of air and moisture should be used to keep the seed dry. During the period immediately after collection, seed is particularly susceptible to diseases and damage so treat it with care. Do not mix seed of different species during collection.

Collecting pods that fall naturally or after shaking the tree

This is the simplest method of collecting seed, and requires very little equipment. A rake, a sieve and a seed container are useful. Care should be taken not to shake the tree so vigorously that unripe pods fall. Seed is mature when the pods are starting to dry or have dried up. Plastic, metal, cloth, or canvas sheets can be laid under selected trees for the pods and seeds to fall on. If these are left for any length of time, they must be checked regularly to prevent losses from birds, mice and insects.

Seed is available to anyone who will collect it

Collecting pods from the ground is best done early in the morning before livestock or wild animals can eat them. Animals often spend the day lying under the trees enjoying the shade while eating fallen pods.

Collecting pods from standing trees

If the seeds or pods cannot be collected from the ground or the parts of the tree that are in reach, then the tree may have to be climbed. Generally, it is not recommended that branches be removed to collect seed, as it is likely that pods which do not fall to the ground when the tree is shaken will contain unripe seed. However, in some circumstances saws or pangas, or a range of secateurs, can be used to remove pod-bearing branches to make collection easier.

A string or small rope with a stone attached to one end can be thrown over branches which can then be brought into reach or broken down by pulling on the two ends of the rope.

Collecting pods from felled trees

This method is cheap and easy and can provide large quantities of seed. However, seed collected in this way usually has very poor viability and this method is NOT recommended.

Crushed pods can be winnowed to remove the seeds

Extracting seed from the pods

After collection, pods can be spread out in the sun to quicken the drying process. Do not expose the seed to naked flames. The seed should now be extracted from the pods.

The pods may be placed in a mortar, then pounded lightly with a pestle. Avoid heavy pounding, as this can damage the seed. Alternatively, place the pods in a sack and beat it to extract the seed. The seed can be separated from the chaff by winnowing, using traditional trays.

Number of seeds per kilogram

It is useful to know the average number of seeds per kilogram. This will assist in estimating the weight of seed required to raise a desired number of seedlings. The expected number of seeds per kilogram for the acacias is given below, although there can be significant variation between seed from different trees and locations.

Table 6.1

Species	Seeds per kg
Acacia erioloba	3 000
Acacia galpinii	3 600
Acacia karroo	18 000
Acacia nigrescens	9 000
Acacia nilotica	5 000
Acacia polyacantha	20 000
Acacia rehmanniana	8 000
Acacia sieberiana	10 000
Acacia tortilis	26 000
Dichrostachys cinerea	48 000
Faidherbia albida	7 000

Source: Zimbabwe Forestry Commission

Packaging and storing seed

After collecting and extracting seeds from the pods, it is important that they are suitably stored to maintain their viability for as long as possible.

Store only quality seed. Two simple means of rejecting inferior seed are:

1. Look at the seed and reject any seed that is small, misshapen, or darker or lighter than usual. Also throw out any seeds that have been holed by insects.
2. Place the seed in water and throw away any seeds that float. The good seed will sink. Pour off the water and dry the seed well before it is stored.

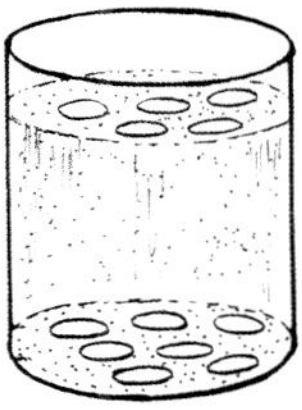

The simplest way to test seed is to drop it in water: If the seed floats, throw it out

The seeds must be kept clean, cool and dry. All acacias have hard seed coats, which gives them the ability to retain their viability for a long time even in storage. Fresh seeds do not generally need the same degree of pre-treatment as older seeds.

Many containers can be used to store seed

Packing and storing seed

Fresh seed that has not been dried should not be stored in airtight containers as it may rot. Sow the seed as soon as possible, as fresh seed has had less time to develop any problems that might affect germination.

Acacia seed should be packaged in dry containers. Small quantities of seed can be stored in jars, glazed clay pots, plastic boxes, cloth, or plastic bags. Plastic bags are convenient because they are easily available, but they should only be used for short-term storage. When storing large volumes of seed, metal or plastic cans or drums can be used. Seed should be stored in a dark place to reduce seed deterioration through exposure to light. Seed can be stored at room temperature, or in granaries where the temperature is lower than outside.

Seed boring insects, especially bruchids, can destroy large quantities of stored acacia seed. The seed should be observed for minute holes in them that indicate an infestation. Any infected seed should be destroyed, as some bruchids can reproduce within the stored seed. Traditional methods of controlling bruchids include mixing the seed with the sawdust of *Spirostachys africana* (mutovhoti), or the ash of *Combretum* (mugodo) species. Alternatively, the leaves of either *Tagetes minuta* (mbanda, chimbanje) or trees such as neem can be mixed with the seed. Bruchid activity can also be minimised if the seed is stored at about 0°C. Chemical seed dressings are effective but expensive.

It is helpful to label all stored seed. Record the species, the date and the place where the seed was collected, and the name of the collector in each package. Attach a duplicate label to the outside of the container.

Seed testing

Seed can be kept on a damp cloth to test germination

To test the quality of seed, take a dish, a piece of cloth, water and find a warm place that will not be disturbed. Count out 25, 50, or 100 seeds and do any pre-treatment required for planting. Spread the seeds out on a damp cloth that has been placed in a dish or bowl. Cover the seeds with a fold of the cloth; moisten it thoroughly and place in a warm (not hot) place such as under a tin roof; the water must not cover the seeds. Keep the cloth moist and check the seeds daily for signs of germination. While germination continues, keep a record of the day and numbers that germinate. The test will usually be complete in less than four days if the seed has been nicked and soaked (see seed pre-treatment on page 48) Simple multiplication provides a germination percentage.

The appearance of the radicle shows that germination has occurred

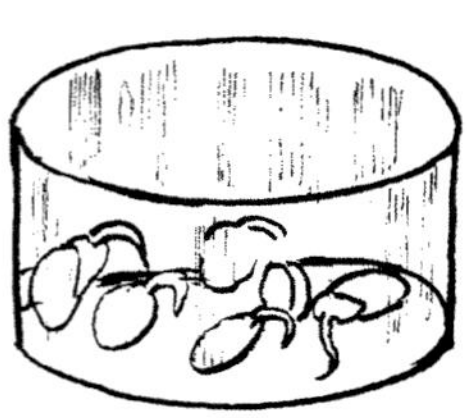

The nursery

The nursery must be established near a permanent water source in an area that is protected from livestock.

Nurseries must be protected and located near a water source

The nursery site should be clean and clear of weeds. A north-facing and gently sloping site that allows water to drain easily is ideal. Areas that are subject to extreme weather conditions like frost, hail, or strong winds should be avoided. If this is not possible, it may be necessary to erect a barrier to protect the seedlings. Grass fences are useful in frost-prone areas.

The nursery should be located close to sources of soil that will be used for filling the seedling containers. Nurseries are often established in corners or along the fences of existing vegetable gardens because they are protected and near a permanent water source.

Soil mixtures

The ideal sowing medium is a free-draining workable of good texture such as a sandy loam. The soil should be rich in nutrients. Usually, a mix of soils provides the best medium for healthy plant growth. An ideal mixture gets its bulk from an inactive, free-draining material such as washed river sand and a small amount of a rich material such as humus or compost.

Nursery soil: Collect topsoil from underneath established acacias

A suitable potting soil can be obtained from mixing one part river sand to four parts of compost.

Another recommended mixture for a good planting soil is made from:

- 1 part topsoil from under neath established acacias
- 2 parts river sand
- 2 parts well-rotted manure or compost (preferably in powder form)
- 5 parts light sandy clay loam (topsoil)

Nursery soil: Collect sand from a riverbed

Nursery soil: Collect well-rotted manure from cattle pens or fields

The soil mixture at every nursery is different because of what is available locally. Aim for a reasonably rich and well-drained soil.

Soil collection should be done as early as May.

The soil should be heaped at a convenient place in the nursery and away from where the nursery beds are to be laid out. The soil is best sieved to remove any coarse particles.

Mixing is done during the sieving process. The texture of the soil should be such that if a ball is made when the soil is wet, it should break easily when hit with a finger. This mixture gives a well-balanced nursery soil that does not compact to create a hard pan at the base of the pot or on the surface as a result of frequent watering.

Nursery soil: Mix the available soils

Rhizobia: Acacias grow in a symbiotic relationship with a bacterium called rhizobium. These rhizobia help leguminous plants like acacias to fix atmospheric nitrogen in the soil.

If you are unable to obtain commercially produced rhizobia, then soil from under naturally growing acacia trees should be added (10% of the total volume) to the soil mixture. This will introduce rhizobia that might otherwise be absent from the normal nursery soil.

The soil can be watered a few times before planting as this will encourage any weed seeds to germinate. The weeds can then be pulled out or dug into the soil.

Nursery containers and soil filling

Acacia seeds are best sown directly into tubes, bags, or other containers. If they are grown in beds, they should be transplanted 7 to 14 days after germination. The seedlings are best grown in long containers because they produce long taproots that should be restricted as little as possible by the

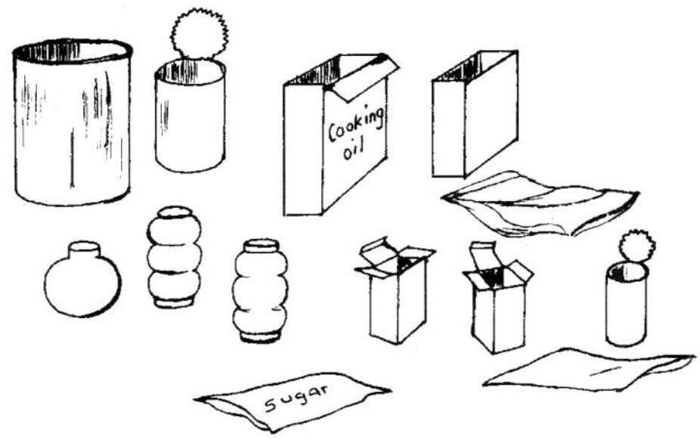

Seedlings can be grown in a variety of containers

length of the container. Black polythene tubes that are 10 cm wide x 25 cm tall (flat measurements) are the most suitable. If they are not available, a range of packaging can be used. This depends on what is available, and can include empty plastic bags, tins, and old pots.

Fill the containers with the mixed soil

Preference is sometimes given to smaller tubes or containers because they require less soil and occupy less space. This cuts down transport costs of both soil and plants. Smaller pots require careful management to ensure that they do not dry out and that the roots are not permitted to grow through the bottom of the container into the ground below. Containers with closed ends should have holes made in them to allow easy drainage, as too much water encourages disease and negatively affects growth. The holes also discourage the roots from coiling up inside the pots. Root coiling in the container can permanently affect the growth of the tree.

Stacking soil around the pots can conserve moisture

Open-ended containers are rammed with moist soil to form a soil plug of about 5 cm at its base to prevent soil from falling out. More soil is then added, compacting it lightly until the container is full. A planting trowel, a funnel, or bare hands can be used for this operation. After filling, the pots can be arranged into beds that are 10 pots wide (for easier stock counting) in rows separated by a path about 50 cm wide. An edge can be placed around the beds in the form of planks, bricks, round timber or stones, or alternatively the seedlings can be dug into the soil. This supports the seedlings, conserves water in the pots and provides the pots with protection from the sun. The pots are watered periodically to allow the soil to settle. Any weeds that germinate can be pulled out before the seed is sown.

DATE OF SOWING AND SEED PRE-TREATMENT

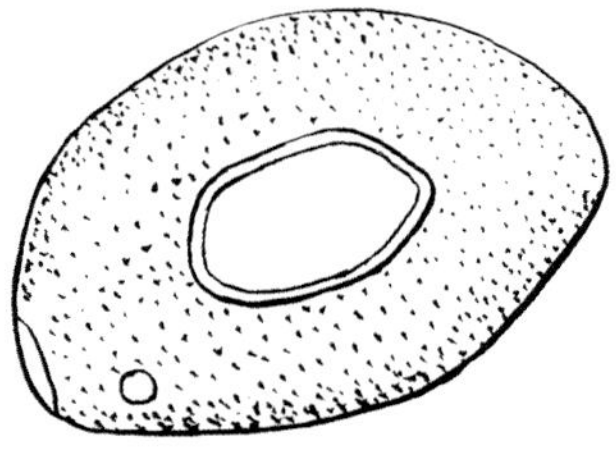

An acacia seed: note the indentation on the left, this is the micropyle

Seed sowing can start from the end of July after the danger of frost has passed. This gives the seedlings time to grow into sturdy plants ready for planting with the rains at the start of December. To raise acacias, it is important to know the basic seed characteristics.

Every acacia seed has a very hard seed coat that needs to be broken, chipped, or otherwise scarified to make it more permeable to water and air. This short-ens the germination period and ensures that the seedlings germinate at roughly the same time. If the seed is not scarified some of it will still not have germinated while others will have already reached a suitable size for planting out. Scarifying the seed involves filing the seed, or nicking it using a nail cutter or a very sharp knife. This should be done at the end of the seed that has a small hole called a micropyle in it. Care must be taken to ensure that the seed coat is chipped or cracked without causing any damage to the embryo inside. Keep the scarification nicks, cuts, or scratches to the minimum size possible. Excessive damage to the testa is a potential site of infection. These methods are suitable when dealing with domestic quantities of seed.

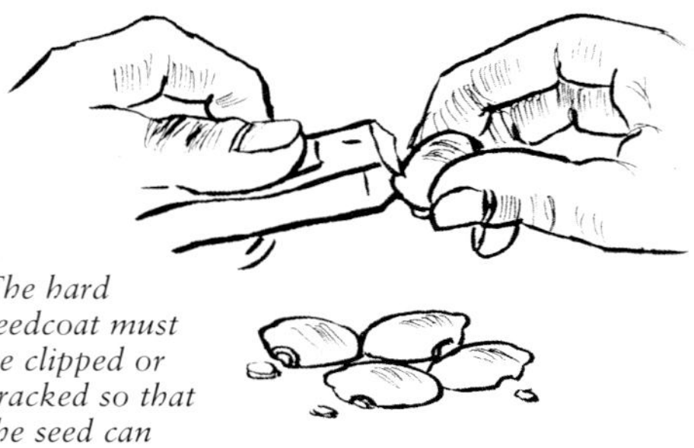

The hard seedcoat must be clipped or cracked so that the seed can take in water

When dealing with large quantities of seed, it can be mixed with river sand and carefully rubbed or pounded in a mortar with a pestle to thin or crack the coat.

After scarification, soak the seed in water for 30 minutes, drain off the water, allow to aerate and dry for an hour or so. Repeat this process until all seeds look well swollen, then plant. An alternative method of treating the seed is to place it briefly into boiling water and then to transfer the seed rapidly into cold water.

NATURAL SEED SCARIFICATION

There are natural processes that can scarify seed and make it germinate easily. For example, when animals feed on acacia pods, acids scarify the seeds as they pass though the stomach. The seed then germinates in the field when it rains. Although there is no way to select the parent trees unless the seeds are collected and fed to the cattle first, dung from animals that feed on pods can be a good source of scarified acacia seed or germinating seedlings. These can be pricked out and put into pots. Bush fires also have the effect of cracking the seed coat, and seed toasted in this way is likely to germinate freely when it rains

A more expensive and dangerous method that is used to scarify acacia seed is to place the seed for two minutes into concentrated sulphuric acid, then to wash it repeatedly and soak it as described above. Unless this process is undertaken by someone with experience it is not recommended because of the dangers involved in working with acid.

Sowing

Seed is best sown after the last frost of the season. A small depression can be made in the soil, to twice the depth of the seed length. The seed is placed in this depression, covered with a thin layer of sand or sieved potting soil, and then watered. Take care not to over-water the soil before the seed has fully germinated. Most losses are due to excess water allowing fungal activity to infect unsprouted seed. Germination should take from 3 to 18 days for most of the species. Shading the plants after germination should be avoided, as they will become spindly and difficult to transplant.

Watering and weeding

Seedlings need regular watering during the first three months of growth

Plants generally need watering daily or at least every second day, depending how hot it is and how quickly the soil dries out, for the first month after they germinate. Gentle but thorough watering is best. A fine rose should be used for watering so that the force of the water does not move the soil. After a month, the seedlings should only be watered when the soil has dried out and the seedlings are starting to show signs of stress. Then they should be given a thorough soaking. This should be followed by another dry period and a subsequent soaking. This pattern of watering and allowing the seedlings to dry out should be continued, always taking care that they do not dry out so much that it causes the seedlings to suffer or die. The best time to water is late in the afternoon. This avoids excessive moisture loss during the hottest part of the day. Watering depends on the weather conditions and the soil structure. More frequent watering is needed in hot weather and in sandier soils than in cooler weather and in loamy soil. Over-watering is a

Weeding is necessary to avoid stunting the seedlings

common cause of seedling losses and should be avoided as it leads to damping-off disease of the seedling. Weeding should be done periodically by pulling out any weeds. Some people find that a small stick can help with this exercise.

Root pruning

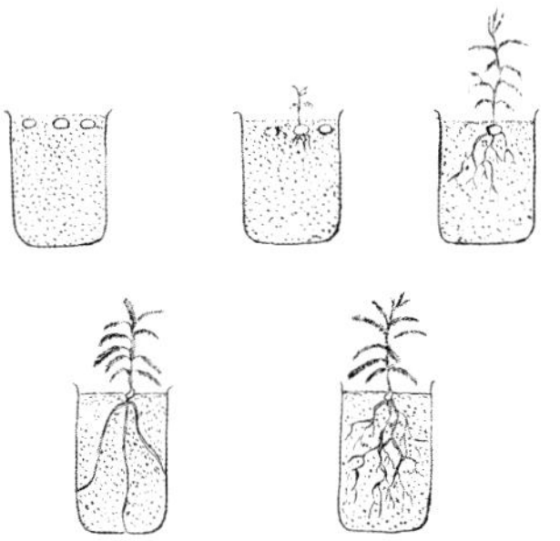

Germinating seed in a pot

As the plants grow, the roots will reach the bottom of the pot. When this happens and the roots emerge from the bottom of the pot, they should be cut off. This action is called root pruning and is important if healthy seedlings are to be produced. Root pruning is normally necessary for acacia species about six weeks after sowing, although this will depend on the size of the planting container.

Any roots that grow out of the bottom of the containers must be cut to encourage a dense root system in the pot

Plants tend to wilt after root pruning and should be watered after the operation. Root pruning is done manually by lifting each pot and cutting any roots growing out the bottom with a knife, pruning shears, or a pair of secateurs. Root pruning encourages a vigorous root system to develop within the pot. If the seedlings are kept in the pots for too long the taproots will coil around the bottom. This will stunt the plant when it is planted out. The pots can be kept on raised wire mesh platforms so that the roots that emerge are 'air-pruned' rather than cut off.

Pest and disease control

Acacia seedlings have very few pests and diseases, but watch out for grasshoppers and caterpillars. These can be removed by hand and crushed underfoot. If mice are causing any damage to the seedlings they should be trapped. Where there is a shortage of surface water, large tins or drums can be dug into the ground and partly filled with water. Smearing a little peanut butter 3 cm below the rim of the drum entices rats and mice to reach for the bait and fall into the water and drown. There are a variety of effective traps and best is to use what is locally available.

Grasshoppers can be removed by hand and crushed underfoot

The most common disease in the nursery is damping-off. This is a fungal disease that is identifiable because the seedlings start to rot from the base of the stem. If this is a problem, immediately reduce the frequency of watering and assure that there is free air circulation to the nursery.

Mice should be trapped

Hardening off

Seedlings should be hardened off or toughened up before planting to prepare them for the harsh conditions they will face in the field.

Root pruning is one way of hardening off seedlings. Another hardening off method is to reduce watering frequency. This can start about six weeks before planting out and involves reducing watering so that the soil dries out and the seedlings start to show signs of wilting. Watering is resumed for a few days until the seedlings recover. This process should be repeated several times. Water the plants well the day before they are taken from the nursery for planting. Great care should be taken when handling seedlings. They should be handled by the pot and not by the plant itself.

Seedlings should always be treated gently

Bruchid Beetles

Seed collected from many acacia species will have been parasitized by beetles of the Bruchid family. The larvae of these beetles feed inside the seed and when it emerges, it drills a round hole in the seed coat. The presence of these exit holes usually means that the embryo has been damaged and that the seed is no longer viable.

There are more than a thousand species in the bruchid family. The adult beetle varies in length from 1 to 30 mm long. They are major pests of legume seed and are sometimes called pea or bean weevils, although they are not strictly weevils. One group of bruchids lays its eggs on the young green pods. The larvae soon hatch and burrow immediately into the seed. Here, they create a chamber and eat away the inner contents until only a thin shell of seedcoat remains. When fully grown, they pupate and the adult beetle emerges after the pod has matured by boring a round exit hole through the seed coat. These bruchid species do not re-infest the seeds after emergence. Another group of bruchids is able to produce many generations on the same stored seed until the food supply is exhausted. The larvae of some species are too large to pupate in the seed and these emerge and lower themselves on a thin silken thread to pupate in the ground.

Many species of the family Bruchidae parasitize seeds of the acacias. 14 of the 61Acacia species in East Africa are known to be hosts to one or more species of bruchid, and the true proportion is likely to be much higher. For example, one bruchid species found in Zimbabwe parasitizes *Acacia tortilis, A. erioloba, A. karroo* and *A. abyssinica*, plus other species outside the country. Predation can destroy all the seed crop, although high rates are normally around 60 to 70%. However, the severity of bruchid attack can vary greatly from tree to tree. This might be due to the flowering time of the individual tree occurring at the same time as the height of bruchid activity; or it may be due to some trees having higher concentrations of chemicals that are toxic to the beetle.

When seed is eaten by cattle, the acids in the gut are thought to kill any bruchids and no further damage is done after the seed emerges in the dung. Insecticides can be used to treat the seed and prevent re-infestation. Freezing can also kill bruchids within the seed before they have finished feeding.

Bruchids do have some value, however; they have been used in the biological control of some species of Australian acacias that have become serious invaders in exotic situations; that is, the beetles have been released on the trees to minimize the viable seed crop.

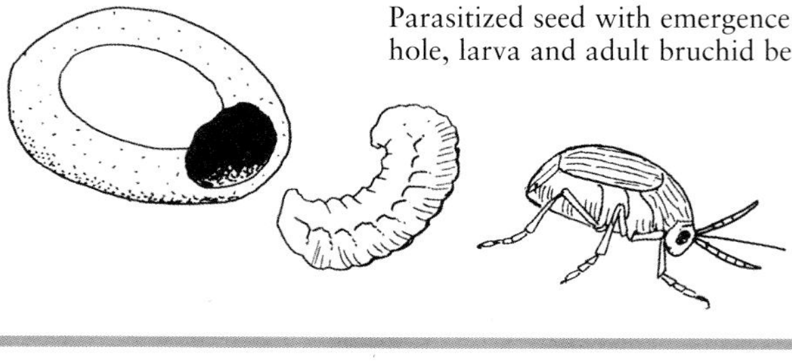

Parasitized seed with emergence hole, larva and adult bruchid beetle

Acacias: Germination and early growth of different species

This section provides summarised information on seed pre-treatment and growth of selected species. All acacias are intolerant of weed competition and benefit from clean weeding during their first season.

Acacia erioloba

Bruchid infestation:	Liable to heavy infestation
Expected germination percentage:	Up to 95%
Recommended pre-treatment:	The seeds have a hard, thick and impervious seed coat, which requires grinding or filing away two edges adjacent to the hilum. If the seed is filed on one side only, the radicle may not be able to free itself from the seed coat. This may result in the death of the seedling. Place scarified seed in water to soak
Time from sowing to emergence:	9-17 days
Growth rate in the nursery:	Early growth in pots is slow and height rarely exceeds 28 cm at planting time. The species tends to produce a substantial root system and seedlings should not be left in the nursery for a long time as they rapidly become stunted
Establishment in the field:	Seedlings do not transplant well
Growth rate during first season:	Slow
Field characteristics:	Once established, the species is tolerant of frost and is very resistant to drought and termites. It is very susceptible to fire but does coppice. Growth rate speeds up considerably after four years in the field

Acacia galpinii

Bruchid infestation:	Common
Expected viability:	Viability is often very low, around 50%
Recommended pre-treatment:	Seed should be notched at the edge, placed in boiling water that has been taken off the heat and soaked
Time from sowing to emergence:	5-8 days
Growth rate in the nursery:	Growth rate in pots is somewhat slow initially but tends to be faster with time
Establishment in the field:	Watering will speed seedling growth. Although fairly tolerant of frost, a severe frost will kill back much of the young growth. Coppices well when young
Growth rate during first season:	Fairly fast
Field characteristics:	Grows into large trees

Acacia karroo

Bruchid infestation:	Often a high rate of bruchid infestation
Expected germination:	Can be very good. Up to 70%
Recommended pre-treatment:	Seed of *Acacia karroo* is difficult to collect as the pods open on the tree. The seed coat is not very hard and nicking or filing can easily break dormancy. The seed is rather small so handling whilst filing can be difficult. Pounding seed in a mortar can provide an alternative pre-treatment method. An alternative is to pour boiling water over the seeds, add cold water and allow them to soak
Time from sowing to emergence:	5-13 days
Growth rate in the nursery:	Growth is relatively fast with seedlings often reaching 50 cm by planting time
Establishment in the field:	Plants are fast-growing and readily establish themselves on degraded sites
Growth rate during first season:	Fast. Responds well to watering
Field characteristics:	Species is resistant to both frost and drought

Acacia nigrescens

Bruchid infestation:	Some infection can be expected
Expected germination:	Over 60 %
Recommended pre-treatment:	The seed coat is not very tough. Filing on one end should be enough to achieve good germination. There is moderate germination without pre-treatment. Radicle growth is fast and where seed is sown in seedbeds, transplanting should be done 7-10 days after germination
Time from sowing to emergence:	Seed can germinate less than 24 hours after sowing and cotyledons should appear in 3-7 days
Growth rate in the nursery:	Seedlings should be about 25 cm in height at planting time
Establishment in the field:	Seedlings respond well to supplementary watering but can tolerate water stress. Susceptible to frost when young, but as the seedlings grow their tolerance increases
Growth rate during first season:	Slow
Field characteristics:	The species is long-lived and is resistant to both fire and drought

Acacia nilotica

Bruchid infestation:	Bruchid beetles seldom attack the seed and viable quantities of seed can be collected quite easily
Expected germination:	Variable, up to 95 %

Recommended pre-treatment:	The seed coat is fairly thick and should be filed and the seed soaked
Time from sowing to emergence:	7-18 days
Growth rate in the nursery:	Fairly slow
Establishment in the field:	The species is susceptible to frost and should be protected. Young plants coppice well. Generally good survival
Growth rate during first season:	Plants are initially slow growing
Field characteristics:	Tolerant to drought once established

ACACIA POLYACANTHA

Bruchid infestation:	Usually a large proportion of the seed is insect free
Expected germination:	High rates of germination (80%) can be achieved
Recommended pre-treatment:	Filing, placing in boiling hot water that has been taken off the heat and left to soak
Time from sowing to emergence:	Seedlings appear in 8-10 days
Growth rate in the nursery:	Growth rate is fair and seedlings can reach 30 cm at planting time
Establishment in the field:	Easily established, especially on heavier soils. Seedlings are frost tender, but become more tolerant as they grow
Growth rate during first season:	Fast. Respond well to watering
Field characteristics:	Species tends to invade old fields, especially those on fertile red clays

ACACIA REHMANNIANA

Bruchid infestation:	Seed is often infected and may need sorting
Expected germination:	Treated seed should give over 50% viability
Recommended pre-treatment:	Seed coat is substantial. Treatment requires filing on one edge to expose the cotyledons. Seed should then be placed in water and soaked
Time from sowing to emergence:	7-11 days
Growth rate in the nursery:	Initial growth is slow, but seedlings can reach 40 cm during the regular nursery period. Seedlings should not be kept for extended periods in the nursery because the roots coil up in the containers
Establishment in the field:	Species is frost hardy, but will be killed back in the first 1-3 winters by severe frost
Growth rate during first season:	Fairly slow. Seedlings respond well to supplementary watering
Field characteristics:	Repeatedly coppices from the root stock

ACACIA SIEBERIANA

Bruchid infestation:	Not common
Expected germination:	70 to 95%
Recommended pre-treatment:	Although the pods are difficult to open, a large percentage of the seed can be extracted undamaged. Normal pre-treatment methods such as filing, placed in boiling water and soaking achieve good results
Time from sowing to emergence:	7-16 days
Growth rate in the nursery:	This is one of the fastest growing acacias and seedlings may reach 50 cm at planting. The seedlings are sensitive to water stress and should be given generous watering to achieve fast growth rates
Establishment in the field:	Sometimes difficult to establish. This could be because some ecotypes are not suitable under certain conditions. Seedlings do show some resistance to frost
Growth rate during first season:	Normally fairly fast
Field characteristics:	Species is commonly found growing in the highveld

ACACIA TORTILIS

Bruchid infestation:	Fairly common
Expected germination:	Very variable
Recommended pre-treatment:	Seeds are small and have a thick seed coat or testa. The process of filing can be tedious. One means of making this easier is to briefly immerse the seed in boiling water before nicking the seed coat. The seed is then soaked until swollen
Time from sowing to emergence:	Germination is fast with radical emergence in 4-10 days
Growth rate in the nursery:	Growth rate is rather slow
Establishment in the field:	Seedlings are tolerant of severe water stress. Susceptible to frost but coppices well
Growth rate during first season:	Often rather slow
Field characteristics:	Establishes readily on degraded sites

DICHROSTACHYS CINEREA

Bruchid infestation:	Not known
Expected germination:	70-95%
Recommended pre-treatment:	Seed is very small and nicking or filing can be tedious
Time from sowing to emergence:	Fairly fast
Growth rate in the nursery:	Fairly slow
Establishment in the field:	Readily established, especially on degraded sites

Growth rate during first season:	Fairly slow
Field characteristics:	Seedlings are common in grazing lands and abandoned fields. Can form thickets

FAIDHERBIA ALBIDA

Bruchid infestation:	Some parasitisation can be expected but good seed in quantity can easily be collected
Expected germination:	Germination up to 95%
Recommended pre-treatment:	Seed coat is quite substantial, but can be scarified easily. After nicking one edge, or briefly immersing in boiling water, the seed should be soaked before sowing
Time from sowing to emergence:	Seedlings should emerge after 7-15 days. Cotyledons are purplish in colour, turning green later
Growth rate in the nursery:	Seedlings can reach 35 cm before planting
Establishment in the field:	The seedlings respond well to supplementary watering in the early stages. Young seedlings are frost tender.
Growth rate during first season:	Initially fairly slow, but fast after the tap root reaches water
Field characteristics:	Plants can reach a height of 7 m in 3 years. Young trees coppice easily

OFFICIAL SOURCES OF SEED IN THE REGION

Botswana	Botswana National Tree Seed Centre, P. Bag 033, Gaborone
Lesotho	Lesotho National Tree Seed Centre, Ministry of Agriculture, PO Box 77, Maseru
Malawi	Malawi National Tree Seed Centre, Forest Research Institute, PO Box 270, Zomba
Mozambique	CEF-Forest Research Centre, Box 8, Marracuene, Maputo
Namibia	SADC Tree Centre Network, PO Box 50295, Windhoek
South Africa	South Africa National Tree Seed Centre, Box 727, Pretoria 0001
Swaziland	Swaziland National Tree Seed Centre, Malkerns Research Station, Box 4, Malkerns, Mbabane
Tanzania	National Tree Seed Programme, PO Box 373, Morogoro
Zambia	Zambia National Tree Seed Centre, Forestry Department, PO Box 22099, Obote, Kitwe
Zimbabwe	National Tree Seed Centre, Forest Research Centre, PO Box HG595, Highlands, Harare

6 ESTABLISHING ACACIAS

Planting seedlings into an unproductive site

Many people believe they do not have enough land to plant trees. They are usually wrong, for even in the most intensively cultivated and heavily populated areas there are places where trees can be grown. This will increase farm productivity. Care during establishment, which is discussed in this chapter, will give the tree the chance of rapid and uninterrupted growth.

PLANNING AVOIDS WASTING TIME AND RESOURCES

The establishment of acacias is no different from that of other trees. As acacias only become productive after some years, careful planning is needed before planting.

Visit the planting site. Work out how many seedlings of each species are needed. Make sure that there will be enough seedlings to plant at the start of the rains, and include an additional 5 or 10% more seedlings to replace any that die in the first year after planting.

By selecting the right species for the site and managing them wisely they can improve site productivity.

The planting process includes planning and involves land preparation, planting seeds or seedlings where they are going to grow, and caring for them for one or two seasons. After this, the surviving trees can be said to be established and can be managed for their intended purpose. Management practices are covered in Chapter 7.

LAND PREPARATION

Land preparation is an important aspect of tree planting and it includes:

SPOT PLANTING

This involves digging and loosening the soil at each planting station. The most effective way of doing this is to dig a hole, 70 to 80 cm deep and 70 to 80 cm wide. The size of the hole depends on the resources that are available. Generally, it is

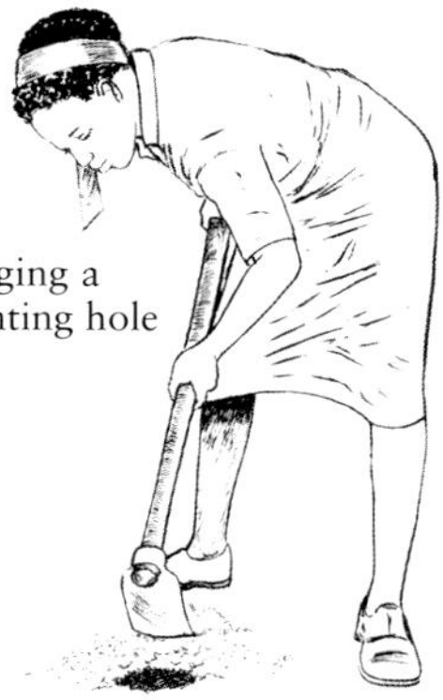

Digging a planting hole

accepted that the bigger the hole, the better the growth. Holes of 25 by 40 cm are suitable for bulk planting of most acacia seedlings. While digging the hole, it is desirable, but not essential, to place the topsoil and the sub-soil in separate piles next to it. When refilling the holes, use the sub-soil first and then the topsoil. If possible, add up to half a wheelbarrow of compost or well-rotted kraal manure to the topsoil and 200 grams of phosphate to both the topsoil and the subsoil to encourage early vigorous growth of the young seedling. Stamp the soil down when backfilling and make sure that the root collar is at the same depth in the soil after planting out as it was in the container.

Line planting

This practice is used when planting seedlings for windbreaks, hedges and boundary markers. Ploughing between three and six furrows along the planting line provides effective land preparation, although planting stations may have to be deepened if the ploughing is too shallow.

Open 3 to 6 furrows when establishing close-spaced seedlings for a windbreak or a live fence

Complete cultivation

This is normal when planting blocks of trees. Usually, the land is cleared. After stumping and removing the cut material, the area is ploughed. Preparatory cultivation can be done in March or April, while the soil is still soft enough to allow easy working at the end of the previous rains and marking of planting stations. Final preparation, with harrowing or hoeing out of weeds and marking of planting stations, can be carried out between October and early November.

Complete cultivation is often done when establishing blocks of trees

The planting site should be prepared early as this allows the planting of the seedlings to be a very quick operation that does not interfere with regular crop planting.

Seedling care before planting

If the seedlings are being brought to the site, make sure that they are available in good time before they are needed for planting.

Seedlings should be well watered the day before transportation.

Whatever form of transport is used, it is important that the ball of soil around the seedling's roots is disturbed as little as possible. Moist hessian sacks are useful for packing round the seedlings so as to keep them from being squashed. The seedlings should be watered again after the journey. Seedlings that are carried in the back of an open truck for any distance are often damaged by the wind so that they lose their leaves and either grow slowly or die. They should be covered with a plastic sheet or shade cloth during transportation to minimise this damage. Old bread crates are useful for transporting and they can be stacked on top of each other if the seedlings are not too tall.

Once the seedlings have been delivered to the farm, they should be placed in a site where they can be watered and protected from livestock until they are planted out. After the initial watering when the seedlings arrive on the farm, they should only be watered when the soil around the roots starts to dry out.

Seedlings are often damaged if they are transported in open vehicles

Seedlings being transported should be covered with plastic sheeting or shade cloth

Planting out

Selected species are planted at or just before the start of the rainy season. Protection from livestock, water harvesting, supplementary watering and mulching can increase survival and growth rates. Tree planting can be integrated with the selective cutting and protection of existing plants at the planting site.

Planting can start any time from October if supplementary watering is possible. Early planting is rewarded by vigorous growth when the rains start and up to two or three times the growth of seedlings planted later in the season. Pre-rain planting is particularly risky in drier areas. If early planting is not possible, it should start as soon as the rains have wet the soil to a depth of 30 cm. This normally occurs by early December, although it can be earlier. Planting should stop if there is a long period without rain. Seedlings should be watered if there is an extended dry spell during the establishment period. Acacia seedlings grow vigorously in direct sunlight. Planting acacias into shaded sites MUST be avoided as the seedlings will fail to thrive.

One way of removing a polythene pocket

Tear or cut the container off so that the soil is not disturbed

Water the seedlings thoroughly before planting. The seedlings should be up to 30 cm tall when planted out, although it is more important for the seedlings to have a well-developed root system than to be very tall. A hole large enough to hold the root ball of soil in the seedling container should be dug at the planting site.

When trees are to be planted into individual pits, which is called spot planting, a hole large enough to hold the root ball is opened in the previously dug and backfilled planting hole. The seedling container is removed to expose the ball of soil. It is important to disturb the soil as little as possible. Polythene tubes can be removed by rolling the tube between the hands to loosen it from the soil and then pulling the tube over the top of the plant. Pockets or other containers with a bottom can be removed by rolling the pocket between the hands. Turning the container upside down with two fingers of one hand on either side of the stem of the seedling, taking hold of

Open a hole large enough for the soil around the seedling at the prepared planting site

the bottom of the container with the other hand and gently pulling it off. If the containers will not come off easily in this way slit it along its length with a sharp knife. This allows the removal of the container without disturbing the soil around the roots.

The seedling is then placed into the previously dug hole and held in position while backfilling is undertaken. The root collar should be at the same depth in the ground as it was in the soil in the container. If there are sufficient seedlings, more than one seedling can be planted at each spot. Mortality will reduce the duplication, and any obviously weaker partners can be removed after one or two seasons.

Plant the seedling with the root collar at the same depth as it was in the container

Firm the soil down to remove any air pockets

Unless the seedlings have been inoculated, or have had soil added from under established acacias to the potting soil, rhizobia can be added at this time to the planting site. It is important NOT to plant the seedlings too shallow or too deep. The soil around the seedling can then be compacted by stamping on it with both feet.

Half-moon water harvesting banks can be built up from the soil.

Containers must never be left on the seedling at planting. Empty containers should be collected and either re-used or disposed of, possibly by burying, or by using as a sort of firelighter.

On sloping ground, open-ended depressions can be dug around the seedlings to trap water. Soil can be built up to channel water to the seedlings. In flat areas a dish-shaped depression that can hold water should be formed from the raised soil around the seedling.

Trees planted in rows are best planted across the slope to reduce soil erosion.

To survive the following dry season, the seedlings roots need to reach a depth where there is permanent moisture. This is most likely to occur when seedlings are planted early in the rainy season in an exposed position that is kept weeded.

Direct seeding

Direct seeding involves planting the seed directly where the tree is going to grow. The seed should be pre-treated. It can either be planted into a prepared site or broadcast after the first rains. Broadcasting seed generally results in a very low seedling survival rate. It is better to broadcast bigger seeds, such as *Acacia erioloba*, as they have more stored resources with which to establish themselves. Small seeds quickly deplete their resources and fail to establish. Direct seeding is not as effective as planting established seedlings and, except under unusually favourable conditions, has a high risk of failure.

Scarification and distribution by livestock

Another means of direct seeding is to feed whole pods of selected acacias to livestock before herding them into the area where it is intended to establish the trees. The seeds are scarified in the animal's gut and a large proportion are then passed out in their dung. The dung provides favourable conditions for germination. Dung containing germinating seedlings can be collected and placed in places where the seedlings are wanted. Although sheep grind up the seed with their teeth, a few do pass through undamaged.

Watering

Try to give planted seedlings a good start. Supplementary watering during dry spells increases plant growth and keeps seedlings alive. This is most essential during the establishment period. Mulch around the seedling will reduce evaporation and assist water infiltration. The mulch will preferably be made of organic material, but where this is not possible stones and rocks can be used. Make sure that the stem of the seedling is kept clear of the mulch by a gap of between 60 and 100 mm.

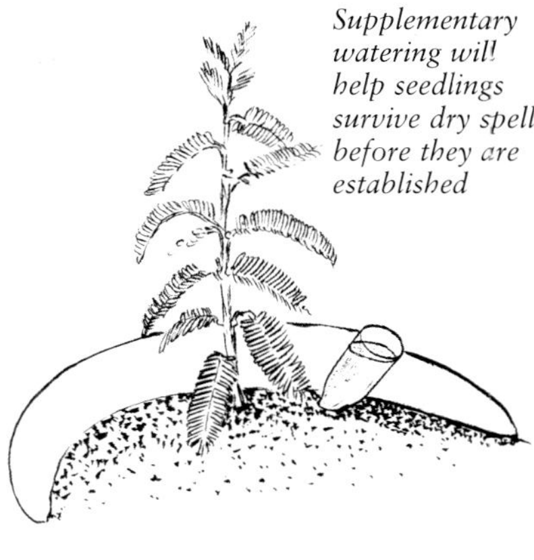

Supplementary watering will help seedlings survive dry spells before they are established

Protection from livestock

Domestic and wild animals eat the shoots of most acacia species. If seedlings are browsed before they are well established, it can cause delays to root development and reduce tree vigour. This may lead to the seedling's death. Protection from animals during the early stages of growth is very important for successful tree establishment.

Seedlings normally need protection

Some acacia seedlings, such as *Acacia tortilis*, can coppice into scrub if they are repeatedly browsed to ground level. When acacias grow older, their thorns harden and provide natural protection from browsing animals. The tree's thorns form a protective barrier around the green shoots and each year the plant gains a little height.

Trees can be protected from browsing in several ways:

1. Fences can be erected around planted areas.
2. Branches of mature acacias can be cut and put over the top of the seedlings. This makes it difficult for animals to browse the newly-established seedlings. If acacias are not already growing, other species' branches can be used, although these are seldom as effective. It is important not to shade the seedlings, or they will not grow well.

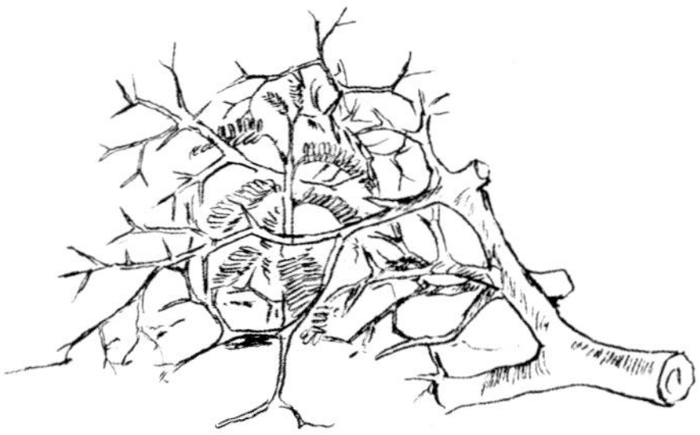

Acacia branches can be placed over seedlings to protect them

3. Smearing dung slurry on planted seedlings is very effective. In termite-infested areas, however, this can lead to termite attacks.
4. Livestock can be herded away from the area after it has been planted.
5. Additional local methods of protecting seedlings should be investigated.

Brushwood fences can be underplanted with acacia seedlings

Weed control/firebreaks

Clear weeds around seedlings

Unless newly-planted seedlings are protected from weeds, they will not thrive. Weeding helps seedlings to survive. A circle with a radius of 40 cm from the seedling should be kept free from weeds until the tree is established. This should be so for at least a year, and it will make significant difference to the seedlings' survival and growth rate. With vigorous grasses, it may be necessary to repeat this weeding a number of times during the growing season.

Firebreaks are recommended in any situation where fire is a danger.

Blanking

Blanking is the practice of replacing any seedlings that die during establishment. It is not generally advisable to plant seedlings after the end of February, unless it is a very wet year or unless supplementary watering can be undertaken.

Planting in specific sites

Homesteads

Trees planted around homesteads can be established using domestic wastewater.

Arable fields

Seedlings planted in fields should be planted early in the rainy season so that they can establish during the time that livestock are herded away from the cropped area.

Contours:

The trees are laid out along the contour line. They can be spaced at 2.5 m intervals, and later thinned as they grow and require more space to spread. The trees can be planted either above, along or below the contour bank.

It is generally agreed that planting halfway between the top of the ridge and the bottom of the furrow is best.

When planting on contours, plant the seedlings halfway between the top of the ridge and the bottom of the furrow

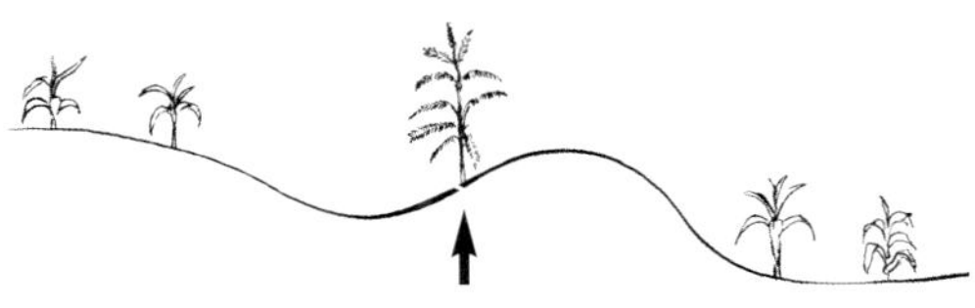

Planting on a contour

Within field

Planting in fields can be done as an intercrop or in a block. Such blocks are usually established in the corners of the field that are seldom cropped. Mixed intercropping involves trees scattered within arable lands. The trees can be laid out at wide spacing along contour lines to ease crop management practices such as ploughing. They can be established at a 5 m by 5 m spacing and subsequently thinned to 10 m by 10 m, depending on management decisions.

Marking planting sites with a rope

Tree fallow cropping

The trees may be established in stands of just one species or in mixed stands. A standard spacing of 2.5 m by 2.5 m or 1600 plants a hectare can be used. In drier areas, the spacing can be widened to about 3 m by 3 m, or about 1100 plants per hectare.

A chain or long rope can be used for line marking. Knots are tied or marks are made at equal distances along its length. The rope is tensioned along the slope. The planting positions can then be marked by digging, or with sticks or stones at the pre-marked points indicated on the rope. After marking the points, the rope is then moved to the next row, parallel to the first, and the process is repeated.

Intercropping plantation trees with annual crops such as maize, beans, cowpeas or vegetables should be undertaken during the first two years that the trees are being established. This will increase production and improve soil conservation in the plantation. Rows should be aligned from east to west to minimise shading of the crop by the trees. Regular pruning of the lower branches and the reduction of forked branches lengthens the period in which intecropping can be successfully carried out. Fodder legumes can also be established in new plantations. It is important to ensure that livestock are excluded from the area until the trees are established. There is a need for high labour input during establishment.

Gardens

These are generally well-protected areas with access to water all year round. Planting can be of individual trees, or blocks, or live fences.

Degraded non-arable sites

Once the necessary support has been enlisted from the traditional leadership, the first step (like the arable areas) should be to survey the area:

- A walking or ground survey is essential to ensure detailed information
- A sketch should be completed of the area, mapping out topography, soils, rocky areas, vegetation, roads, water points and structures
- On the sketch, map out sites that are suitable for tree establishment

As the sites are degraded, water retention, absorption and the reduction of run-off and evaporation will assist establishment. These can be increased by forming contours, mulches, micro-catchments and the like.

After planting, the biggest problem is to protect the seedlings from livestock. If fencing is not available, cut acacia branches can protect newly planted seedlings. Other forms of tree guard or other means of protecting seedlings may be known locally. Retaining patches of bushy thorn scrub in an area provides protection to tree and grass seed. This allows them to germinate and establish out of the reach of browsing and trampling livestock.

However, once established, young acacia trees are able to protect themselves to some extent by their thorns.

The seedlings should also be protected from fire. Many species are fire tolerant and will regenerate after a fire, especially in the early years of seedling growth. However, if the leading shoot is damaged or destroyed then poor and irregular stems and trunks will result which will reduce productivity.

Communities themselves must take responsibility for ensuring that fires are kept out of their woodland but the establishment of a fireguard can be particularly helpful. A two to three metre strip should be cleared around the different woodland blocks. This is to prevent fire entering the woodland. If a fire does enter the woodland, then it assists people fighting the fire to contain it within a block. Grazing can reduce the grass cover, and with it the fire hazard.

River valleys

When planting acacias in river valleys, trees planted near livestock watering points require particular protection to avoid trampling and browsing.

Planting for particular purposes

Fodder

Trees can be grown individually or in blocks. They generally need to be protected. The trees can be grown in arable areas for use during the dry season.

Acacias can be cut and fed to livestock.

Live fences

Live fence poles.

Live poles can be established by planting seedlings at about two metres along the fence line. Careful pruning can encourage the formation of stems to replace conventional poles. Wire can then be strung between the stems. The stems can be pollarded above the level of the top wire to harvest fuel and fodder while retaining the function of the live pole.

Hedges and windbreaks

Hedges produce a wall of vegetation. They can be planted either in single row or double row layouts.

A single row layout consists of one row of plants with a spacing of 25 to 70 cm within the row. Planting should be at the closer spacing in higher rainfall areas and at the wider spacing in lower rainfall areas

Acacia hedges can provide effective barriers against livestock

A two-row layout consists of a double row of plants; the two rows being offset against the other. Select an inter-row spacing of 25, 50 or 75 cm, and an intra-row spacing of 25, 50 or 75 cm, depending on the rainfall.

A large number of seedlings is required to establish hedges and windbreaks.

It is often desirable to combine different species to establish an effective hedge. The different growth form of available species can be combined so that they complement each other in providing a livestock-proof barrier. Local hedge species should be identified and may include *Dovyalis caffra*, *flacourtia indica*, *Portulaca species*, *Ziziphus mauritiana* (in the lowveld), citrus (in wet areas) and any local species of *ficus*. Gaps in the fence can be filled in with plants such as sisal *(Agave americana)*, cut branches (particularly acacias) or barbed wire.

SIX guidelines for successful tree planting

- Plant the right tree on the right site
- Choose the right species for your ecozone
- Plant strong healthy seedlings of the right size
- Plant early, either from October to November if supplementary watering is possible, or as soon as the soil is moist to a depth of 30 cm
- Minimise root competition by weeding the seedlings
- Protect your seedlings from livestock and fire

7 MANAGING ACACIAS

This chapter covers what a farmer can do once trees have become established. It describes different management techniques and suggests when they should be used. It also gives some guidelines on how to manage acacias to get the most value from them. The techniques used in any particular situation depend on what product or products the farmer wants, and whether the acacias were planted or occurred naturally. Although it may have been decided to grow or manage the trees for one particular reason, the trees will continue to provide multiple products and services.

Acacia trees make valuable contributions to many households. This contribution could be significantly increased if people make the change from just being users of acacia products to becoming managers of acacias. Trees should be managed to maximise the output of desirable products. The aim should always be to harvest trees in a sustainable manner. This allows for their continued growth and development, although they are being regularly harvested.

Some trees just grow by themselves

Management helps to improve the quality of standing trees. Management techniques include thinning, coppicing, pollarding and pruning. A number of products are obtained from these activities. They include poles, firewood and brushwood for fencing, and browse for livestock, although these products may not have been the direct reason for managing the trees. The key to making the best use of acacias is through their management.

The options or techniques used for managing a stand of acacias are not very different from those used to manage any other type of tree or bush. The techniques chosen, however, will depend greatly on what products the farmer wants. Management for livestock fodder production will involve different practices from management for poles or firewood. The various management options involve manipulating tree spacing, height and shape.

Management of existing stands usually differs from management of planted stands. With planted stands a lot of effort has gone into establishment, so careful management is needed in order not to lose that investment. Natural stands, on the other hand, are mostly on poorer-quality land, less investment has been put into them, and individual trees are of differing ages. What is generally required in this case is thinning, shaping of the trees, and the provision of a suitable environment for regeneration.

MANAGEMENT TECHNIQUES

Natural woodlands can vary from poor scrub of limited species to areas with diverse species of various age-groups, including established mature trees. As with all management, a continual process of assessment is needed. Decisions must be made regarding the objectives and the actions that are necessary to achieve them. If the woodland is in poor condition, the first objective may be to improve the quality of the woodland by harvesting selected areas or single trees. All poorly grown, diseased, deformed and over-mature trees can be removed, unless they are to be retained for seed, fruit or shade. This can encourage rapid growth of straight stems of young trees of desirable species. The operation is beneficial to the woodland as it reduces nutrient and water competition between the trees and allows light penetration to the understorey, thereby encouraging grass and seedling growth.

While carrying out management operations, it is important to keep tree damage to a minimum. The use of axes and machetes for cutting often injures the tree and exposes it to fungal infection. A long-handled pruning saw or bow saw makes the job cleaner and easier.

THINNING

Thinning involves the removal of dead, damaged or diseased trees, or healthy trees that are being overcrowded. Their removal gives room for other trees to develop and opens the canopy to allow light to reach the ground, which in turn encourages grass growth.

Thinning gives the remaining stems an opportunity for vigorous growth

Thinning intensity depends on what the trees are being managed for. A normal, moderately dense acacia woodland contains around 500 trees per hectare. Villagers normally thin or clearfell a stand when there is a need for firewood or poles. It is usually desirable to maintain multi-layered woodland. This ensures that animals continue to have access to browse and that a range of desirable products can be obtained.

Thinning can be done to maximise wood production when there is 80% or more crown cover of the ground. This is generally after the first pruning. The intensity of thinning depends on the management objectives, but 25% to 50% stem removal to 250 to 375 stems per hectare is normal. Where the woodland is used for grazing, the canopy can be reduced by about 25%. Where growth is vigorous, up to 50% of the canopy can be removed.

Should the initial tree stocking be less than the desired density, thinning should be restricted. Regeneration or enrichment planting can be encouraged or undertaken until the desired stocking level is reached.

Coppicing

The regrowth or coppicing of felled trees lessens the need for planting after trees are cut down

Coppice is the regrowth from stems cut at, or just above, ground level. Most acacias coppice well, especially when young. Coppice growth is faster than growth from seedlings because the plant already has an established root system. Regrowth of *Acacia karroo* on good soils, for instance, can reach 3 metres in a year.

A thinning exercise will result in coppice regrowth. To permanently reduce the stand density the thinned trees should be stumped, or repeatedly cut back. It is best to cut trees for coppice at the end of the dry season before the rains. At this time, the new coppice grows well. The trees should be cut low down, about 10 cm above the ground. If the cut is made higher, valuable timber is lost, the stool does not coppice so well and the coppice shoots are more likely to break off in high winds. A sloping cut should be made so that rain-water flows off the cut surface. Care must be taken to ensure that the surface is as smooth as possible, because if water is trapped on the top of the stool it tends to rot.

After cutting, a number of shoots will usually develop from each stool. During the flush of growth a year later, it is advisable to select two or three of the strongest shoots from each stool and to carefully remove the rest. The shoots that are to be retained should be vigorous and should not be growing together on the same side of the stool.

POLLARDING

Pollarding involves cutting the main stem or branches at a particular height above the ground, usually 1 to 2 m or higher. Regrowth occurs at the point where the stem or branches are cut. Pollarding encourages the development of a large and branching crown. This is useful for improving fodder availability for livestock, for harvesting brushwood fencing and for producing small sized fuelwood. The cut can either be made within the reach of animals or above the animal browse line so that direct access to the browse is restricted.

Careful pollarding can raise the regrowth of fodder above the reach of livestock

Pod production is related to crown size and tree maturity. Old branches generally produce more fruit than new branches. Continuous pollarding will decrease the number of flowers, thereby limiting pod production.

PRUNING

Pruning benefits the tree by trimming out unwanted or unhealthy plant growth. The objective is to increase productivity and to improve stem form. It is done for the following reasons:

- to remove dead, diseased or injured stems or branches
- to alter the form or size of the plants, this may be to thin crowns that have become too dense to admit light and air, or where interior branches do not receive enough sunlight, or to allow livestock to graze the grass that grows under the trees, or to encourage fruit, flower or foliage production
- to encourage root growth and reduce dieback of branches

Acacia bushes respond well to selective pruning. One or two of the main shoots are retained and the remainder are re-moved. This encourages the development of a sound straight stem and good height growth. Pruning leads to faster growth since the nutrients available to the plant are channelled to a

One or two main stems are allowed to remain when pruning acacias

crown and stem that are out of the reach of browsing animals. Avoid over-pruning the tree because this will slow plant growth. As a rule, no more than 30% of the crown should be removed.

Basic pruning methods

Make all cuts as close to the stem or branch as possible. Cuts should be made just above a bud. Try to ensure that all cuts are made cleanly without causing the branch to splinter or tear off the bark below the cut. Saws are more effective than axes and pangas. When pruning large trees, pruning should be done from below in order to reduce crown depth.

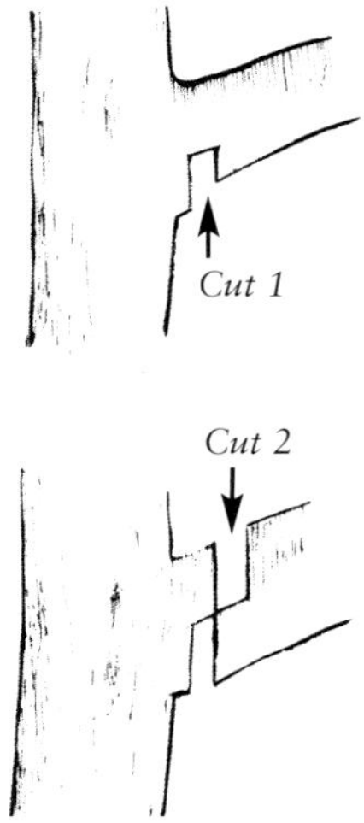

To prevent the bark peeling off when pruning large branches a cut should be made from the bottom of the branch (cut 1) before it is cut through from the top (cut 2)

Pruning large branches

It is important that large branches do not tear the bark from the stem as they are cut. Stems damaged in this way provide an entrance for insects and disease. To reduce the chances of this happening, an undercut should be made on the branch as close to the stem as possible. Saw about one third of the way through the branch. Next, make a cut about 2 cm further from the trunk, from the top of the branch. Continue cutting until the branch falls off.

Timing of pruning

Pruning can be done throughout the year, although it is best done during the dry season before the rains when the tree is dormant or leafless. At this time of year, the framework and branches are easily distinguishable. Pruning of diseased and damaged tree parts should be done as soon as the disease or damage is noticed. Any diseased plant prunings should be burnt.

Management of plantations

Plantation establishment takes a lot of effort and resources. To make this investment worthwhile a high value crop is required. Often the costs of production outweigh the expected returns. It is generally expected that integrating trees into the existing farm activities and landscape will provide better returns.

Plantations can be established for the production of gum, timber, poles or for soil improvement. They consist of trees of the same age. The planned product will dictate the stocking density and management routines. A dense stocking rate is suitable for firewood, but a much more open stand is required if good quality timber is required. Pruning should be carried out regularly. Thinning will usually be carried out after two to three years to remove weak plants and to reduce the stand to the appropriate stocking density. This may be followed by coppicing.

MANAGEMENT OF NATURAL STANDS

Often a farm has an area of naturally regenerating woodland. On suitable soils, this may consist of various acacia species and the area involved can be quite large. If the land is not immediately needed, the farmer should look on this as a good opportunity. Here is a plantation without any of the costs of raising seed or establishment. What is required for good production is only to manage the stand. Suitable management of acacia woodland is more important and valuable than is usually recognised. With management, what is often regarded as bush, of little use other than grazing or 'thorn trees to be got rid of' can become a major asset on a farm.

Erioloba parkland

Firstly, the farmer should decide what is the most useful product that could be obtained. This could be browse and pods for livestock, it could be firewood or poles, or it could be general land improvement together with improved grazing.

Secondly, the management techniques chosen will depend on the product required. In many cases, these may include severe thinning, both of the number of individual trees and the number of stems on each plant. Coppicing is important. Some control over potential thickening-up of the vegetation, or bush encroachment, will also be needed.

REGENERATION

An important part of acacia management is to make sure that the trees thrive and can reproduce themselves. Utilisation should be sustainable so that the stand of trees continues to provide goods and services into the future. New recruitment from seed has to be encouraged before stump mortality and lack of coppicing begins to affect overall productivity.

Sustained production means that dead and dying trees need to be replaced. It also means that trees should not be weakened so much that they cannot produce, unless this is done as part of a thinning exercise. It is expected that some stumps will die after each cutting. Additionally some of the trees will naturally die. If the trees die in such numbers that there is less than 80% of the desired stocking rate, then their numbers should be increased to the desired level. The best, and cheapest, way of ensuring replacement of trees is by natural regeneration. A balance can be struck between regeneration from seed and coppice. Acacias regenerate naturally from seed that can lie dormant in the soil for years. Dormancy can be broken following a fire, or through the actions of naturally occurring acids in the soil, or because of extreme changes in temperature or soil moisture. When seed dormancy is broken, germination occurs after it rains.

Mature acacias produce many seeds, but only some of these germinate into seedlings. It is normal for most of these to die within a year. Management will in certain years create suitable conditions for seedlings to flourish. This allows a good number of seedlings to become saplings that are strong enough to withstand browsing, drought and fire so that the next generation of trees can come through.

Root suckering is an effective way to increase the number of trees in an area. It involves cutting the roots and leaving the cut ends exposed above the ground. Shoots sprout from the exposed ends and can grow quickly because the roots are already established.

Root suckering

Most acacias will not grow in the shade. The other main reasons why seedlings die in the first year is because of competition from grass or weeds, from a lack of moisture in the soil, or because animals eat them. Farmers can reduce these threats to encourage regeneration, or they can increase the threats if wishing to control unwanted regeneration. The most important management techniques to encourage seedling growth are to reduce competition from grasses and to protect them from grazing.

Management for particular uses

Although acacias may be grown primarily for one particular product or service, it is through the production of multiple products that optimum benefit can be gained from the trees. Maximum production of different products requires that trees be managed in different ways. While management of acacias can be started at any stage during the growth of the trees, it is most desirable to start when the trees are small or bushy. It is generally true to say that a well-established acacia parkland will provide maximum benefit in both products and services.

Managing acacias for livestock production

Acacia foliage can be fed to livestock

The management of acacias in grazing lands should aim at maximising the production of pods and foliage, while at the same time allowing for good grass growth. The production of these depends on the size of the tree crown. Management should aim at increasing the crown size for fodder, but still allow enough light to reach the ground for grass growth.

Fodder produced by acacias is particularly valuable during the dry season when grass and other fodder is in short supply. Pods provide an important source of protein for livestock, as grass during the winter becomes scarce, dry and less nutritious. Acacias provide their main feed value to livestock between July and November.

For good browse production, mature trees should be widely spaced and should branch low down. This means the leaves remain within reach of livestock. Cattle can reach up to 2 metres, while goats can reach only 1.5 metres. Dense stands should be thinned to allow access for livestock to graze and browse. Planting 1600 stems per hectare of *A. karroo* and thinning these to about 400 plants a hectare is suggested as an optimum plant density for mixed cattle and goat herds. This suggests a spacing of 2.5 m between adjacent plants that are subsequently thinned to 5 m between plants. This spacing can be applied as a guideline for the other species. As the trees mature, alternate trees can be removed to provide more space for those remaining. Mature trees produce better quantities and quality of pods and seeds than younger trees.

Erioloba parkland with tall grass

A density of around 25 mature trees a hectare (20 m between trees) of *A. erioloba* is desirable as this low density allows enough light to get through for good grass growth. In addition, there is adequate shade for livestock. The competition between trees is not very great, so production of pods is substantial. This spacing can be varied as pod production per hectare from large, broad-canopy trees spaced up to 50 metres apart can be enormous.

There is no hard and fast rule to apply in every situation. There is a strong argument to seek a balance of trees with high canopies for shade and pod production while some trees are repeatedly pollarded or coppiced to keep some browse in constant reach of livestock. Some species of acacia, including *Acacia erioloba, Acacia karroo, Acacia tortilis, Acacia nigrescens* and *Dichrostachys cinerea*, facilitate the establishment of nutritious grass species such as *Panicum maximum* under their canopies. However, at high tree densities, acacias can suppress grass growth beneath their canopies. Hence tree and grass growth needs to be balanced.

Trees pruned for timber

This balance can be complemented by sound livestock management involving rotational grazing, which ensures extensive grass regrowth and avoids the moribund build up of dead grass.

Managing acacias for timber

Acacias are generally not particularly good timber species. Sawn planks are often twisted or of poor quality. However, production levels can be surprisingly high in some of the harsher areas. Acacias have an advantage over eucalyptus because many resist termite attack. In addition, the saplings are better able to survive drought and browsing by livestock.

Good timber-producing acacias will be found in a variety of sites around the farm where they have little competition. When managing for timber production, individual trees or small groups of trees showing good form of the lead stem are selected for retention and any other shoots are removed. The objective of timber management is to obtain and maintain good tree form, especially straightness of trunk. This is done by coppicing, thinning out poorly formed individuals and by pruning.

Managing acacias for poles

Only a few of the acacias are good for pole production and these are generally slower-growing than the others. Acacia stems are often too twisted for house construction, although they make good fencing poles. Maximum pole production requires a moderately dense stand with relatively small trunks. For fencing posts the trunks should be straight and about 10 to 15 cm in diameter. For droppers the stems should be around 3 to 5 cm in diameter.

In natural stands of acacias, the trees should be thinned to the same densities recommended for planted trees. Surplus trees are removed. Some trees can be left for large pole production, while others are cut for smaller poles. Production of small poles can be encouraged through coppicing. This way, a regular supply of droppers can be maintained.

Managing for live fences and hedges

The growth form of the hedge must be carefully managed if it is to provide an effective barrier to livestock and to reduce competition with adjacent crops for space, light, moisture and nutrients. The aim is to produce a wall of vegetation from top to bottom. This is achieved by snipping off the tip of the leading shoots after establishment and by regular trimming from the time the seedlings reach a height of about 30 cm. This practice encourages the trees to bush outwards into a dense mesh. The pruned branches can be used to fill any gaps and to protect the new growth from browsing animals. This also reduces competition for space and light with adjacent crops. A hedge that is clipped into a pyramidal shape has the advantage that the sides receive more sunlight and are therefore not so liable to become sparse and straggly near the base. Nutrient and moisture competition with adjacent crops can be reduced by digging a trench to a depth of about 75 cm between the crop and the fence line.

Acacias can be clipped to form inpenetrable hedges around gardens

Tree fallow cropping

Tree fallows are used to regain soil fertility for crops, encourage grass for grazing, and produce fooder, fuelwood and/or poles. The trees are felled when the best combination of these services and products has been reached. There is a high labour input needed during the removal (destumping) of trees.

Managing for shade and ornament

Trees should be pruned when the buds are forming. Pruning increases vegetative growth and subsequent flowering and fruiting. It also provides fuelwood and can give the trees a desired shape.

Managing for gum production

The main indigenous acacia species to produce a commercial quality gum is *Acacia karroo*. In Zimbabwe, the gum collected is exuded through natural lesions generally caused by wood-boring insects. In the Sahelian region of Africa, where *Acacia senegal* trees are managed for gum production, strips of bark are removed from the trees using a sharp tool. There should be no more than three cuts per tree. Gum oozes through the incisions. After one or two days the globules of wet gum that have formed harden and can be collected and left to dry. *Acacia karroo* does not seem to respond well to tapping and it seems that the best way to harvest from these trees is by collecting naturally produced gum.

Gum harvesting

To ensure easy access to the stems during gum collection, it is important to keep the acacia stand clean and protected from fire. Fires damage the bark and introduce impurities like charred bark into the gum.

Gum can be collected from *Acacia karroo* throughout the year but is greatest between August and October.

Gum dissolves in water – if it rains before collection the gum may be washed away. Monkeys and birds feed on the gum and so collecting is sometimes done early in the morning, before they get there. The gum is sorted by hand into lumps, finer material and rejects. Rejects comprise any dark gum. Experienced collectors have few or no rejects.

At present, the market for *Acacia karroo* gum is limited. However, indications are that the commodity will become more important and the financial returns of harvesting far more attractive.

Management and control of acacia invasions

Acacias can become weeds, particularly on abandoned, degraded or deforested land. It is on loamy soils that they are most likely to become invasive. They can form dense impenetrable thickets, or low bushes that cover the ground. This is known as bush encroachment and can severely limit the productivity of an area. As is the case with most weeds, they can be extremely difficult to kill. A lot of money can be spent trying to rid an area of acacias with only limited success. Combinations of control methods are generally more effective than any one practice on its own.

It is important to appreciate that an invasion of acacias onto a piece of land provides the farmer with opportunities as well as problems. Obviously, a field full of acacias is a major problem if a farmer is growing maize in the field. However, if the farmer changes the land use of that area to grazing and wood production, the acacias can be very useful. An area of acacia bush encroachment provides the basis for the establishment of an acacia parkland. Acacia parkland is beneficial to the grasses growing in the area and provides many other products to the farmer.

Goats can be used to help control acacia invasions

If many unwanted seedlings are established, there are various ways of weakening or destroying them. One of the best ways to reduce their numbers is to encourage the growth of the larger trees by pruning and thinning out the smaller stems around them. This encourages grasses which smother young seedlings and leads to productive parkland. The most suitable combination will depend on the type of land, how dense and tall the stand of invading acacias is, and how extensive the stand is.

Controlling acacia encroachment is an ongoing process and not a one-off activity. The process starts with an assessment, and involves pruning, thinning and coppicing, which all provide useful products such as poles and firewood. The branch debris can either be laid randomly over bare ground, laid and compressed on contour lines, or piled in spots to leave free passage in surrounding areas. In addition, browsing, buring, removing, ring-barking and poisoning of trees form part of a farmer's options when controlling an acacia invasion.

Pruning and thinning

It is often suggested that the most cost-effective way of controlling bush encroachment is through a programme of pruning and thinning. This process involves the encouragement of selected bushes to grow into mature trees that overshadow and subsequently cause the remaining bushes to die.

Browsing

Heavy browsing by goats will control the spread of acacias, but will rarely eliminate them.Coppicing will occur when the goats are removed. Goats can be very effective in keeping an invasion under control.

Fire

Fire can be a difficult and dangerous tool to use in controlling bush encroachment. Because of this its use is not generally recommended. A fire can easily get out of control and do considerable damage. More is often destroyed, in terms of value (grazing, wood and browse) than will be improved or created. Great care should be taken if burning is to be done. Before lighting a fire, make sure that all neighbours have been notified, that suitable fireguards are in place and that people, water and beaters are on hand to control it. A hot fire with a wind behind it is exceptionally difficult to control. Burning should only be done in cool, non-windy weather, preferably in the late afternoon or evening.

Burning during the hot dry season will kill the above-ground growth of many standing shrubs if there is plenty of fuel (dry grass and dead wood) but it is also likely to damage the trees that you wish to retain. Although many shrubs will appear to have been killed, most will coppice from the rootstocks. This can result in a dense multi-stemmed thicket that is worse than before. Fire can also scarify seeds in the soil, which germinate and cause additional problems. Seeds in the soil may remain viable for decades and new growth from them can continue to be a problem for many years. Allowing a dense grass cover to grow after a fire, by protecting the area from all grazing, will help to kill young coppice and seedling growth. Alternatively, browsing goats can be used on the regrowth immediately after a fire.

Provided your selected trees are large enough, careful use of a controlled fire can achieve a thinning of lower branches of trees and shrubs. This allows goats or cattle to graze underneath and does minimum damage to the existing trees because of their thick bark.

Remember that fire can cause more problems than it solves.

Its use is not generally recommended.

Removal

Stumping, or physical removal, is effective for *Acacia tortilis*, *Acacia karroo* and *Acacia nilotica*. However, where the trees were mature and had fruited over a number of years there are often many seeds stored in the soil that will continue to

Goats can help control acacia regrowth

germinate for years. *Acacia polyacantha* and *Dichrostachys* are particularly difficult to remove by stumping.

Careful burning afterwards, or heavy grazing by livestock, may be necessary to stop re-invasion. Acacia seedlings do not tolerate heavy shade well. By encouraging a good cover of tall grass for two or three years, most seedlings germinating will fail to establish. The farmer can also plant or encourage the growth of other tree species. These will eventually shade out the acacias and stop re-establishment.

Physical removal is extremely expensive if done over a large area. A farmer must always consider whether the costs are justified in terms of the intended land use. Stumping is unlikely to be justified except in intensive farming systems such as irrigated cropping or in improved pastures.

Ring-barking

Ring-barked tree

If a strip of bark is completely removed from all round the base of a tree then nutrients will not be able to travel above the removed section of bark. The tree will quickly coppice and the problem can reappear.

However, provided any coppice shoots that grow below the point of ring barking are repeatedly removed the tree will die.

Arboricides

These are chemicals that kill plants. They are usually applied to a cut stump, or painted on to a cut stem. These chemicals are very expensive and bush clearance using them is unlikely to be economic. The arboricides also have undesirable environmental effects.

This chapter has looked at some basic management practices that can be applied to acacias. There are additional practices that can be applied in different situations. So look carefully at what is already being done on neighbouring farms or similar areas elsewhere.

CASE STUDY 1: MANAGING ACACIAS FOR POD PRODUCTION

Acacia erioloba pod meal forms an important part of cattle feed at the Umgusa Valley Estates, near Bulawayo in Matabeleland. The pods are collected annually, milled and mixed with stover to provide a livestock feed. Naturally regenerating trees on previously cultivated lands are being managed to create parkland for pod production and improved grazing.

50 trees were selected from a 20-year-old stand of *Acacia erioloba* and 25 pods were collected from each of these trees. The pods were analysed for their nutritional content and the results compared with the values for maize stover and maize grain. The results were as follows:

NUTRITIONAL VALUE OF ACACIA *ERIOLOBA* (pod and seed meal) COMPARED TO MAIZE (stover and grain)

Parameter	*Acacia erioloba*	Maize grain	Maize stover
Crude protein (%)	16.5	9.6	4.2
Crude fibre (%)	27.3	2.3	39.8
Ether extract (%)	6.3	4.9	1.2
Ash (%)	3.8	1.6	7.4
Metabolisable energy (MJkg-1 DM)	9.4	13.3	4.6
Insoluble tannins* *(Absorb.@ 550nm/g NDF)	10.0	0.0	3.2

It can be seen that the crude protein, which is the most expensive part of animal feed, in the acacia pod meal is nearly twice that found in pure maize grain, and four times that found in maize stover. However, only about 75% of the crude protein in the acacia pod meal is digestible, due to the presence of tannins.

The yield from individual trees was recorded to be more than 150 kg of pods per year. Fifteen trees can produce more crude protein than an average smallholder farmer can produce from a hectare of maize grown for livestock feed. There are additional benefits to the farmer as there are no costs of ploughing, weeding, fertilisation or seed incurred when growing trees. The input required for the trees will be the cost of pod collection and milling only. In this situation, it is clearly more profitable to give priority to the production of livestock feed from trees rather than from maize.

To achieve sustained production of pods from parkland, young trees should be protected and brought into production to replace old trees as they become unproductive. The acacias have the added benefit of improving grazing beneath their canopies, as well as providing a desirable microclimate for livestock comfort in harsh environments.

Case Study 2: Ntabazinduna – village-based woodland management

Ntabazinduna is a heavily populated area about 40 kilometres north east of Bulawayo. In the 1960s, the area was deforested when the trees were cut for fuelwood for the city. When a local resident, Mr. Maviva, returned after a few years absence, he was shocked by the bareness of the area that was previously well forested. He began to prune the small bushes, mostly *Acacia karroo*, and to protect them from cutting to allow them to grow. While he and a few other community members protected trees near their homes, the rest of the area become increasingly deforested until only small bushes and limited grass covered the grazing areas. In the area that was pruned the results were impressive. There was greater shade, access for livestock and increased fodder available for both cattle and goats while the pruned materials provided fuel and brushwood to the practitioners.

Mr. Maviva was elected ward councillor in 1983 and immediately began to promote tree management on a wider scale within his ward. By 1984, villages adjacent to his had also started managing the acacia scrub in their grazing area.

The success of this small group in improving the grazing and tree cover gradually began to influence the others in the ward, as well as people in neighbouring wards. The councillors in adjacent wards began to support the practice in their own areas.

Community-based woodland management projects took a major stride forward during the 1991/92 drought, when villagers voted to include pruning and thinning operations as one of the activities under the government's food-for-work relief programme.

Woodland management pilot projects began in 11 of the 12 villages in the West and North Wards in communal grazing areas. The practice of pruning and thinning rather than cutting is now, according to the councillors, spreading on its own.

Pruning and thinning takes place through organised group work, as happened under the government food-for-work relief programmes. It also occurs on a day-to-day basis by individuals who want firewood or brushwood fencing.

Each bush is thinned down to one main stem. Whole plants can be removed if they have poor form or are growing too near each other. This initial pruning allows for an increased growth rate of the remaining stem, as well as improved grass growth between the bushes. Later, as the tree grows, the side branches are also pruned. The cuttings are spread on the bare ground to reduce run-off and to promote grass establishment. As the tree grows, the pruning moves further up the stem until it reaches a height that allows cattle to move freely underneath the canopy.

None of the protected areas are fenced, despite the high stocking rates of both cattle and goats. Livestock pose no threat to the regrowth of the trees. The effect has been to increase the stocking capacity of the area and to increase the production of tree products such as fuelwood and brushwood.

There is legal provision to prosecute those who cut trees without authority under Rural District Council by-laws, but this is seldom used. The councillors estimated that only one case a year is brought before them.The community projects rely on the commitment of all people in a village. This co-operation is a testimony to the patient and tireless work of Mr Maviva who has consistently led by example. His emphasis has always been on raising awareness and love for trees, so that the people enact the law themselves.

8 ACACIAS IN EXTENSION AND TEACHING

This handbook is intended to assist extension workers and teachers of conservation, forestry and agriculture who are responsible for linking appropriate technical information with farmers and the public.

There are numerous people who will be happy to give you assistance with tree planting

It will be necessary for teachers and extension workers to be selective in deciding which material will be appropriate to their particular audience. It may also be necessary for the material to be modified to suit the needs of that audience and the particular locality. Subject areas that can be extracted from the handbook include site selection, species selection, seed collection, seedling production, tree planting and tree management. These are only the most obvious subjects that the manual covers.

The starting point in any training, research or extension activities should be an exercise in participatory analysis. This diagnostic process provides information on the target area and can be based on the Farm Survey and an exercise involving the following:

- Analysing the local farm system
- Prioritising the critical problems
- Looking at opportunities for addressing those problems
- Planning a course of action
- Implementing the plans
- Reflecting on the activity

At the end of this activity an evolutionary cycle of planning, implementation and reflection should be maintained.

Bold initiatives are required to raise awareness of the benefits of planting and managing acacias. We know that there are tremendous pressures on the environment because of population increase and present farming methods. By linking the power of agricultural science with the problems and opportunities that exist within our farming communities, we can begin to reduce these pressures.

The extension worker

There is always a need to consider the social, economic and cultural values of the community when taking decisions about groups planting and managing trees. These include issues such as the use of acacias in burial rituals and local rules regarding prohibition on burning of some acacias. These values are established in oral tradition and the best way to find out what is important is through talking to different people.

When working on environmental issues, it is useful to remember the past. Two unsustainable trends predominate in the region's agricultural development:

- Excessive use of external inputs on large scale farming enterprises which has led to environmental degradation and depletion of non-renewable resources
- Erosive forms of low-external-input agriculture in communal areas, which has resulted in the natural resources no longer being able to support the local people

The issue of tree planting should involve the whole community

Conventional research and extension advice to farmers has been based on the total removal of trees from arable lands, the complete removal of all but a few shade trees from grazing areas, the use of hybrid seed, heavy applications of fertiliser and pesticides, monocropping and an intensification of mechanisation. The negative social and environmental impacts of these practices have become increasingly obvious, particularly in conditions where farmers are forced to over-exploit the resources that are available to them. A cycle of poverty and dependency has been developed. Deterioration in these areas can be slowed, halted and reversed by the adoption of ecologically sound agricultural practices. An important step towards this is the integration of trees into other farming activities.

Appreciation of acacias must be encouraged. Acacias have thorns and because of this, there is often resistance to their use. The benefits of the numerous products and services of acacias far outweigh their prickliness. Any resistance people have to these thorny friends must be overcome. In areas where acacias are the dominant trees, people make good use of them in spite of their thorns. Bold initiatives are required to raise awareness of the benefits of planting and managing acacias. They provide a means to limit some of the negative effects on the environment of population pressure and unsuitable agricultural practices. Farmers must change their position from being destroyers or, at most, just opportune users of acacias, to becoming managers of acacias for their many products and services.

Farm survey

A farm survey helps you to make decisions about what trees to plant, where to plant them and how to manage them. Before starting the exercise a decision must be made as to the detail of the survey.

Generally, a farm survey involves a careful look at your farm to see what it contains, what it may be lacking and what constraints or opportunities exist for development. The farm survey can include one or more of the following activities:

- The completion of a record sheet detailing landscape, rainfall, soil, vegetation and on-farm crops and livestock activities. The **Farm Survey Record Sheet** (page 93) provides a framework for issues to consider when undertaking a farm survey
- A farm sketch should be completed, mapping out fields, topography, soils, rocky areas, vegetation, roads, streams, water points and structures
- A walking survey is essential to ensure that accurate information is detailed during the farm survey

If the survey is being done as the basis for a tree-planting exercise, be sure to include the following information:

- What species of trees are already growing on the farm and in the surrounding areas? Trees that are already growing in an area provide a source of seed and show that they can survive. Give priority to these species as they are already proven.
- Where are they growing on the farm? The planting sites or niches listed on the **Existing Tree Data Sheet** (page 94) provide a framework for recording where trees are growing, or can be grown, on the farm.
- Were the trees planted, or do they occur naturally?
- Have the trees been protected from livestock or not?
- In what types of soil are the trees growing?

THE TEACHER

The teacher has numerous opportunities to use the handbook both inside and outside the classroom. Trees can be brought into various subjects in both primary and secondary school curricula.

- One way is to integrate tree-growing into the Environmental Science curriculum, which is already in place in primary schools. Tree growing and tree care normally fall under the topics "Trees and Forestry" or "Crop Plants and Animals". In some cases, the study of trees relates to the topic "Soil, Grass and Grazing". Trees and tree growing can be an excellent science learning resource. See the syllabus for specific learning activities

- Trees can be used to introduce or add to other subjects such as mathematics, art, language development,writing, and geography. This can be done by involving students in wide-ranging activities, including games, puzzles, songs, plays, riddles and collections that involve trees

THE STUDENT AND THE NATURAL ENVIRONMENT

The study of trees and natural forests can be used as a tool for studying the following categories found in the primary school Environmental Science Syllabus:

ECOLOGICAL SCIENCE CONCEPTS

Photosynthesis, growth, decay, habitat, variety and change

PHYSICAL SCIENCE CONCEPTS

Materials, properties, work, energy

GEOGRAPHICAL CONCEPTS

Shapes, scale, altitude, location, space

ABSTRACT CONCEPTS

Cycle, system, energy, interdependence, balance, health, climate, productivity, resource and conservation

As noted in the syllabus, the Environmental Science programme has been designed to make the practical activities of each grade self-reliant as well as supportive of other grades. This encourages understanding and positive attitudes towards community interdependence. Developing school nurseries and woodlots is an excellent way for classes and grade levels to work together.

Using trees in other subject areas

The following are some ways that the study of trees can be used in other subject areas.

Mathematics activities:

- Count trees in a given area and note the increase or reduction over time. Present the data on a graph
- Measure the area of planting site, woodlot, nursery, leaf, fruit, shadow sizes
- Compare tree shapes and sizes and the sizes of shadows at different times in the day
- Measure distances between trees, or measure branch angles
- Try to calculate the approximate amount of usable wood in a tree
- Measure the height, width, circumference, diameter or radius of tree trunks
- Present data in the form of graphs

Language development

- Name trees, label them, compare their uses and group them by use or name
- Arrange tree names in alphabetical order
- Record seasonal changes of trees in diaries. Have students write and perform speeches, plays, essays, personifications, word games or debates about trees

Social studies/geography

- Study the relationship between climate and trees
- Research local uses for trees and their significance (ecological, economic and religious)
- Interview local people and record the history of the woodlands in the area
- Do population studies of trees

Art

- Woodwork and carving
- Compare textures of tree barks by making rubbings
- Use dyes and fibres obtained from trees
- Compare shapes of leaves, seeds, fruits, etc.
- Draw the shapes and patterns of trees, canopies, leaves, fruits, and seeds
- Use trees as subjects for colour work, such as shading or grouping
- Make collections of items found in and around trees. Look for similarities and differences
- Make collages using a variety of seeds, leaves and bark
- Make puppets of animals that use trees as a habitat

Trees seeds can be used for a variety of art practicals

Other activities

- Involve children in tree-centred activities such as games, puzzles, songs, plays, riddles, language, maths, art, and collections
- Invite guests to speak to students about the different values and uses of trees. These can include representatives from groups such as traditional healers associations, the forestry authority and agricultural extension services
- Make up songs about tree names and characteristics, conservation and values
- Make instruments and trophies from wood
- Arrange a tour to a botanical garden or nursery
- Take the class outside and study particular trees or establish trees at the school, or undertake a community outreach programme

These suggestions are applicable mostly to primary schools, although many secondary school subjects offer similar opportunities. This is particularly true for the following subjects: agriculture, geography, integrated science, biology, maths, art, building and home economics.

Trees enrich our lives

FARM SURVEY RECORD SHEET

Date

Location

Farmer's name

LAND

Rainfall
Vegetation
Size and description of land
Homefield — Mainfield — Garden —
Soils: Indicate differences between fields
Include information on fertility and degradation
Land form, (hillcrest, valley floor, etc.)

GENERAL

Labour availability:
- Family
- Hired
- Other (community)

Secondary activities
Extension services
Seasonal food availability and quality
Fuel use and availability

CROPS

What crops - A. Field crops B. Garden crops
- What activities are required for each crop, and when(include tools)
- Inputs
- Soil fertility maintenance
- Most common rotations
- Most common intercrop combinations
- Protection
- Major problems growing crops

LIVESTOCK

Type and herd structure
- Management
- Feed sources A. Grazing B. Other
- Pests and diseases
- Marketing
- Major problems with livestock

TREE AND SHRUB CROPS

What tree products do people need and want?
Who controls access to trees?
Is there anybody in the community who is raising, managing or planting trees?
Establishment practises: Have you:

Transplanted wild seedlings?	If so, which species?
Collected seed?	Which?
Direct planted seeds?	Which?
Raised seedlings?	If so, which?
Taken cuttings?	If so, which species?
Planted seedlings?	Which? And where?
Had any training in tree growing?	What training?

MANAGEMENT

Are there any sacred trees or woodlands in the area?
What tree management activities are undertaken?
What are the major problems you face planting trees?

Existing tree data

The letters below record if the trees are found:

A. B. C.	A. in ones or twos	B. in small numbers	C. in large numbers
P. U.	P. protected	U. unprotected	
N. PL.	N. naturally occurring	PL. planted	

LOCATION	SPECIES	A. B. C.	P. U.	N. PL.	Soil type
HOMESTEAD:					
HOME FIELDS:					
A. On contours					
B. In the field					
C. On the boundary					
D. Along path:					
GARDENS:					
MAIN FIELDS:					
A. On contours					
B. In the field					
C. On the boundary					
WOODLOTS:					
CATTLE KRAALS:					
BOUNDARIES					
SCHOOLS:					
OTHER:					

A

ACACIA SPECIES DESCRIPTIONS

This chapter provides descriptions, illustrations and other information on the species that are covered in this handbook. These are:

Acacia erioloba	Acacia galpinii
Acacia karroo	Acacia nigrescens
Acacia nilotica	Acacia polyacantha
Acacia rehmanniana	Acacia sieberiana
Acacia tortilis	Dichrostachys cinerea
Faidherbia albida	

Although *Dichrostachys* and *Faidherbia* (which was formerly called *Acacia albida*) are not true acacias, they are included because they share so many characteristics with them.

For each of the species, information is provided on their distinguishing features, their distribution in South Central Africa, their ecology, growth pattern and cultivation. A brief account of the uses of each species is also given, as are their common Zimbabwean names.

Many acacias are found growing naturally in the region. All of them have thorns that are either long and straight, or short and hooked. A full list and identification guide to all 40 species which grow in Zimbabwe can be found in the companion book to this publication, 'FIELD GUIDE TO THE ACACIAS OF ZIMBABWE' (Timberlake *et. al.*, 1999). Similar information for the acacias that grow in South Africa can be found in Smit (1999), in Botswana in Timberlake (1980), in Kenya in Coe and Beentje (1991).

Botanical names are used throughout the handbook to allow common understanding, regardless of what language people use. English, Shona, Ndebele, Afrikaans and other local names are listed on page 135, although they often refer to more than one species. This is well illustrated by the Zimbabwean names commonly used to describe the two broad groups into which acacias can be divided: These groups are:

a) Acacias with long straight thorns and with flowers in yellow or white balls. These are called *isinga* or *umlaladwayi* in Ndebele and *mubayamhondoro* or *muunga* in Shona.
b) Acacias with smaller hooked thorns and with flowers in white spikes or balls are called *umkaya* in Ndebele and *munanga* or *chinanga* in Shona.

Very useful species may have a range of names in different areas, while species of little use or significance are often not individually recognised and are lumped together with others to which they are similar. Minor differences of spellings and pronunciation of vernacular names have not been included. A list of local names from the region is given at the end of the handbook.

In this chapter the sizes of leaves and leaflets are defined as follows:

- Small-sized leaves are those generally less than 3 cm in length long
- Medium-sized leaves are those generally 3-8 cm long
- Large-sized leaves are those generally over 8 cm long
- Small-sized leaflets are those generally less than 4 mm long
- Medium-sized leaflets are those generally 4-11mm long
- Large-sized leaflets are those generally over 11mm long

The usual numbers of leaves, pinnae and leaflets are given.

Identification of species is made easier when pods are present and is aided by the comparison of the leaves.

An explanation of the terms used to describe the leaves, thorns and flowers is given at the end of this chapter on page 130.

ACACIA ERIOLOBA (previously called Acacia giraffae)

COMMON NAMES *umwholo*, *umholo* (Ndebele); *camel thorn*, *giraffe thorn* (English); *kameeldoring* (Afrikaans)

DESCRIPTION

SIZE/SHAPE *Acacia erioloba* is often a large spreading tree, 8-15 metres high. It keeps its bluish-green leaves for much of the year. It usually branches low down.

BARK Deeply fissured and dark brown in colour.

TWIGS/THORN Young twigs are zigzag in shape with pairs of large, strong, straight thorns.

LEAVES Medium-sized, with 2-5 pairs of pinnae and 8-15 pairs of blue-green medium-sized leaflets with glands on the rachis but not on the petiole.

FLOWERS Bright yellow balls that appear with the new leaves early in the season from August to October.

Acacia erioloba

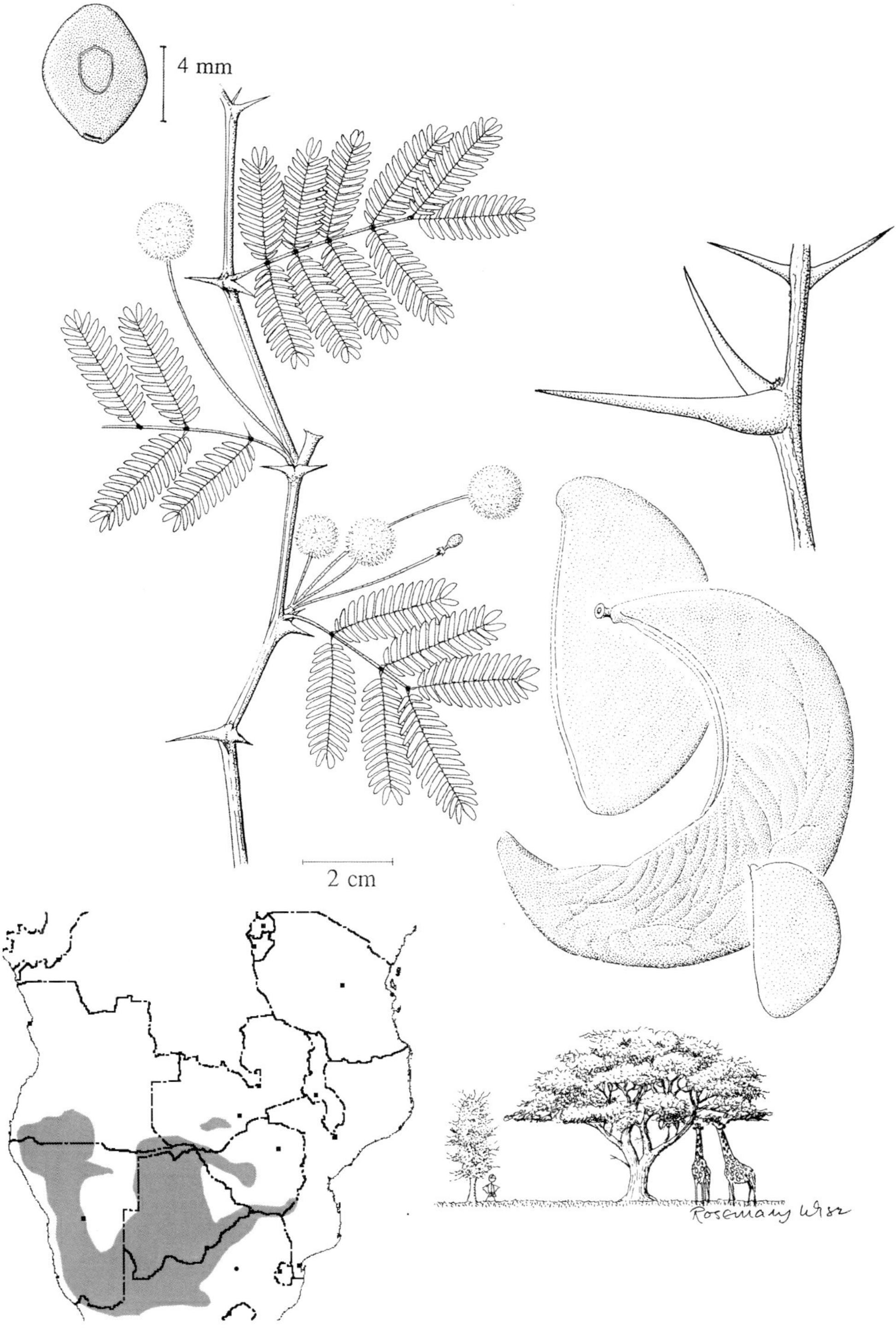

PODS	Large, curved, grey, furry and very distinctive. They are woody and indehiscent. The pods mature from June onwards, although a few are found on the tree all year round.
SEEDS	Large (8-13 x 7 x 10 mm), rounded, with a very hard coat.
FIELD CHARACTERISTICS	Large spreading tree on sandy soils: zigzag shape of young twigs; large straight thorns often with swollen bases; large woody, grey, velvety ear-shaped pods; dark grey-green foliage.
DISTRIBUTION & ECOLOGY	This species is confined to southern Africa, where it is found from southern Angola and Namibia across to western Zimbabwe and northern South Africa. It also just enters south-western Zambia. *Acacia erioloba* is very typical of the Kalahari sandveld and is an important part of the landscapes of Namibia, Botswana and the northern Cape of South Africa. Across its range it occurs on Kalahari sand or other loose fine sandy soils, generally in places with slightly more soil nutrients such as old river valleys, around pans and in dune slacks.
GROWTH & CULTIVATION	*Acacia erioloba* grows very slowly for the first 4-5 years while it establishes deep roots. After the roots are established, its growth is quite fast and it can reach 8 m in 20 years if it is not heavily browsed. Large specimens can be over 100 years old. It is deep rooting and does not transplant well. Commonly found on sandy soils, it does not generally grow well on clay or saline soils. Trees will usually grow better if calcium is applied at planting. The species is tolerant of frost and is very drought-resistant once established. It coppices well when cut back and regenerates readily from root suckers. The trees are resistant to termites. Wild animals can damage young trees.
USES	Firewood and pods are the two main usable products, but the smoke has a distinctive smell which can be unpleasant in an enclosed space. The wood is generally too twisted and prone to splitting to be of much use for making furniture. It has been used for machine bearings owing to its hardness. It provides very strong posts, which are used for fencing or rough construction. The dark red heartwood is durable and resistant to fungi, borer and termites. It is heavy and dense and takes a high polish. The bark has been used for tanning in Zambia.

The major economic value of *Acacia erioloba* comes from the large pods it produces. These provide valuable fodder for cattle and wildlife, especially during the dry season. The milled pods contain 10-16% crude protein, which is more than maize-grain contains. Production can range from 500-2000 kilograms per hectare per year. Individual large trees can produce 500 kilograms of pods a year. Pods are best collected and crushed or milled to break open the protein-rich seeds and then fed to cattle. If they are not crushed, the seeds pass through the cattle without being digested. Some farmers add a little sulphur to neutralise a toxin that is sometimes present in younger pods. The seeds have also been used as a coffee substitute.

A valuable fodder is also provided by the green leaves, which contain up to 17% protein. The flowers are relished by livestock but are not visited by bees. It is an important shade tree over much of the Kalahari region.

There are a number of medicinal uses reported: powdered pods are used to treat ear infections; the bark is burnt to ease headaches; and the edible gum is used to treat sexually transmitted diseases (STDs). *Acacia erioloba* is an indicator of well-drained sandy soil with good depth.

ACACIA ERIOLOBA	
Use – production:	timber, firewood, fodder (pods), medicine
Use – service:	shade, soil improvement
Growth rate:	initially slow
Coppicing ability:	good
Environmental tolerances:	drought, frost, fires
Compatibility with arable agriculture:	moderate to good
Soil type:	sandy
Soil fertility improvement:	good

Acacia galpinii

Common names	*Umdwadwa*, *umtungabayeni* (Ndebele); *nkotokua*, *mfututa*, *munanga* (Shona); *kalaunga* (Shangaan); *monkey thorn* (English); *apiesdoring* (Afrikaans)
Description	
Size/shape	*Acacia galpinii* is a large tree up to 30 metres high when mature. It has an erect branching habit when young, but the crown becomes rounded with age. It is leafless for much of the dry season.
Bark	Greyish to grey-brown and often flaky on very young trees.
Twigs/thorns	The large thorns are paired, hooked and blackish in colour. Some persist on the branches and trunk, where they are situated on small woody knobs.
Leaves	Large-sized, generally with 9-14 pairs of pinnae and 13-35 pairs of medium-sized leaflets. A small raised gland is seen in the lower half of the petiole.
Flowers	These appear before the leaves very early in the season around August/September, forming long cream-coloured spikes, tinged reddish before they open.
Pods	Long (10-25 cm) and broad, but fairly brittle and thin. They are light brown when young, sometimes with a purplish tinge, and darker brown when older.
Seeds	Large (12-15 x 10-12 mm), almost circular, but flattened and olive-brown in colour.
Field characteristics	Tall, magnificent tree usually found on riverbanks or termite mounds; big, loose crown when mature; sometimes with a flaky bark when young; long thin spikes of cream flowers with reddish tinge before the leaves come out.
Distribution & ecology	This southern African species is found in southern Zambia, north and eastern Botswana, Zimbabwe, northern South Africa, Malawi and eastern Mozambique, and also in central Tanzania. *Acacia galpinii* is generally found in cleared areas on deeper loamy soils where it can be a pioneer species. It is moderately deep-rooted and is more common in drier areas at medium altitudes. Typically it is associated with drainage lines and can become a very large tree along some rivers. It is also found on termite mounds.

ACACIA GALPINII

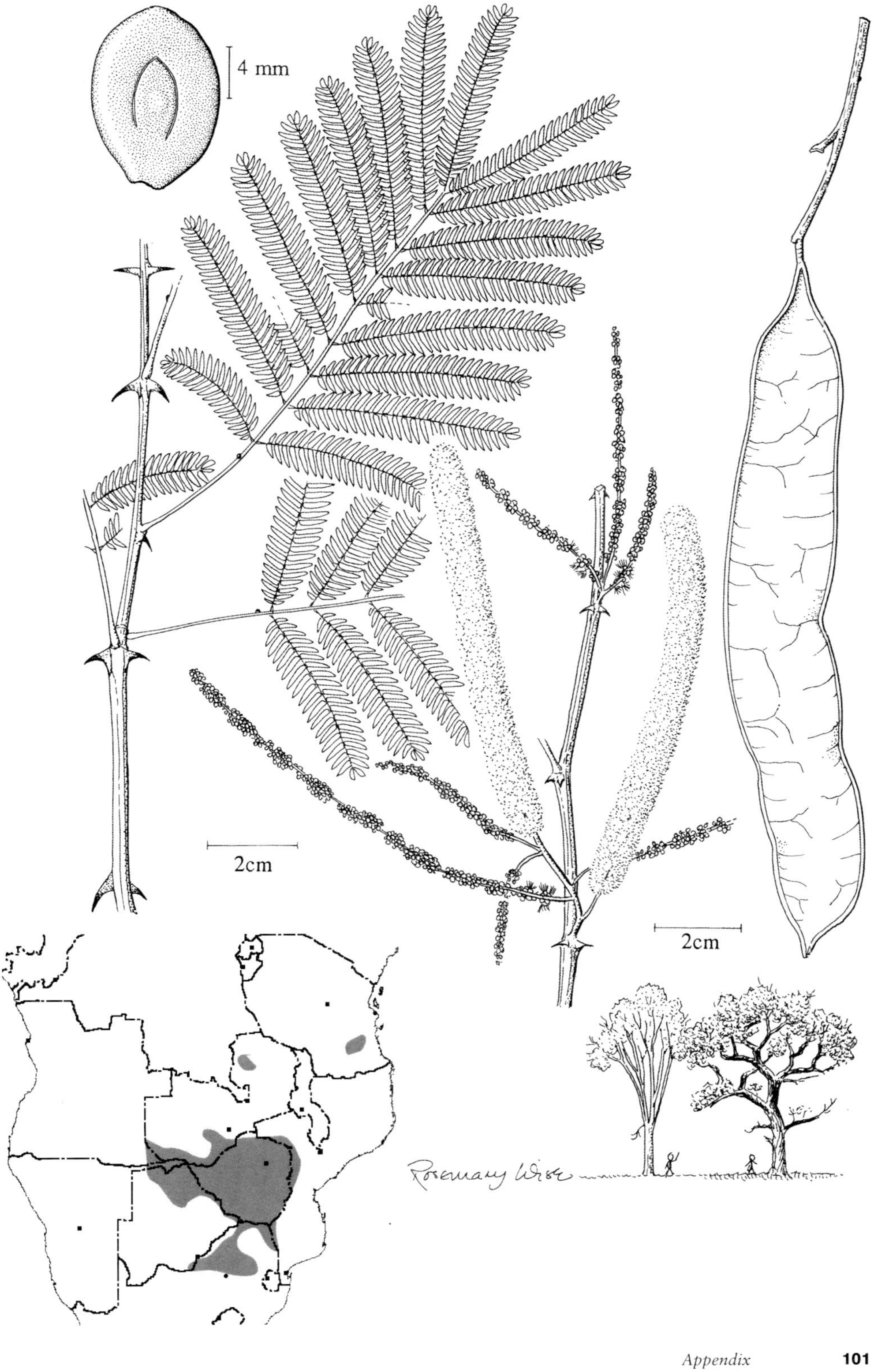

GROWTH & CULTIVATION

Acacia galpinii is easy to germinate and establish. It is often planted in urban areas as it grows fast. Seedlings respond well to watering during dry spells. Trees coppice well when young although a severe frost will kill back much of the younger growth. Individual trees grow to 50 years of age or more.

USES

The wood is of good quality for furniture, rough construction and planks; it also makes good firewood. It gives the highest timber yield of acacias (up to 2.5 cubic metres per tree from large trees). The hard, heavy heartwood is durable and very resistant to termites or borers. It has also been used for fencing and carts. Both livestock and wildlife eat the pods. The flowers provide good nectar and pollen.

ACACIA GALPINII	
Use – Production:	timber, firewood, and bees
Use – Service:	shade
Growth rate:	very fast
Coppicing ability:	reasonably well when young
Environmental tolerances:	some frost tolerance
Compatibility with arable agriculture:	moderate
Soil type:	clay, loam
Soil fertility improvement:	moderate

ACACIA KARROO

COMMON NAMES

isinga (Ndebele); *munenje*, *mubayamhondoro* (Shona); *butema*, *gaba* (Shona-Kalanga); *sweet thorn*, *mimosa thorn* (English); *soetdoring* (Afrikaans)

DESCRIPTION

SIZE/SHAPE

Acacia karroo is a medium sized tree usually up to 8 metres high, but it can reach 17 metres. Its branches reach up and form a rounded canopy, although individuals in crowded stands can be spindly with many dead branches. It retains its leaves through part of the dry season.

BARK

Dark brown to black on the trunk, rough when old but smoother with a yellowish to orange tinge when young.

TWIGS/THORNS

Bark on twigs often flakes off to expose a reddish underlayer. The thorns are paired, long, straight and whitish in colour. They are sometimes slightly swollen along their length.

LEAVES

Medium-sized, with 2-7 pairs of pinnae and 5-15 pairs of shiny green medium-sized leaflets. The leaf stalk has a shallow furrow along its length.

Acacia karroo

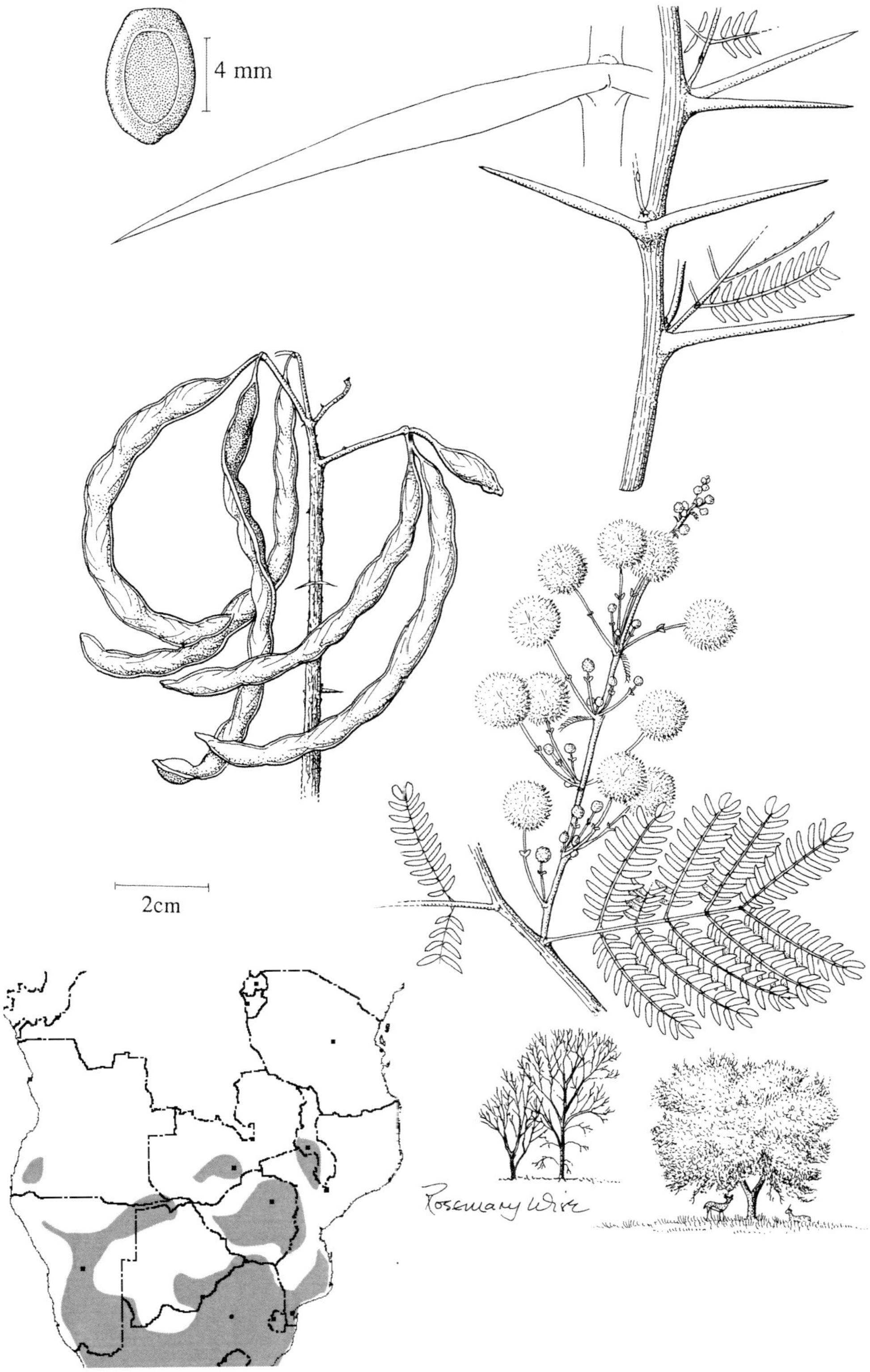

FLOWERS	Grow in golden-yellow balls at the ends of twigs. They appear after spells of heavy rainfall from November through to January. There can be two or three flowerings a season, depending on the rainfall pattern.
PODS	Clustered, narrow (around 0.5 cm wide), long and curved. They open on the tree from March onwards. When open, they are often twisted.
SEEDS	Small to medium (5-8 x 3-5mm), elliptic or rounded, they hang from a thread and are olive-brown in colour.
FIELD CHARACTERISTICS	Medium-sized, upright tree along water-courses and on black cotton soils; rounded crown with curved ascending branches; bark almost black on trunk, rusty brown or orange on branches; glossy rich green leaflets; bunches of slender sickle-shaped pods. Profuse golden-yellow balls of flowers from November to March.
DISTRIBUTION & ECOLOGY	This is perhaps the most widespread acacia species in southern Africa. It occurs from southern Malawi across to southern Angola and southwards to the South African and Mozambican coasts, but is absent from the dry interior of Botswana. It is a species of open woodland and bushland, and is often found in association with other acacia species. Across its range it is very variable with a number of different growth forms. In Zimbabwe, *Acacia karroo* is principally found in the highveld above 1000m altitude, generally on clay-rich soils where there is access to soil moisture for much of the year or along watercourses. Elsewhere it is found in a range of habitats from the arid Karoo to the sand dunes of the KwaZulu-Natal and Mozambican coasts.
GROWTH & CULTIVATION	*Acacia karroo* is a fast-growing pioneer species which readily establishes itself in suitable disturbed areas. Many overgrazed areas or old lands throughout its distribution are now occupied by stands of various ages. It is not a long-lived species and often starts dying at around 25 years old. Few trees are likely to be older than 30-40 years. The species is resistant to both frost and drought once established and coppices readily when young. It is sensitive to competition for moisture in its early stages. The seeds germinate easily. For successful establishment the seedlings must not be shaded by other trees or tall grass. They can tolerate slightly saline soils. Stem form can be improved by pruning.
USES	*Acacia karroo* has many uses including wood, gum and live-stock forage. Although the species is fast-growing, the wood is quite heavy. It does not produce much heartwood and it is not particularly durable as it is rapidly attacked by borers and

fungi and is therefore not suitable for outdoor use. The wood is tough and resilient but liable to split and twist. It has been used for furniture and rough carpentry. Although it is good for firewood, producing much heat with little smoke, it does not produce long-lasting coals. It does, however, make good charcoal.

The inner bark has been used to make rope and baskets and for tanning leather as it contains up to 19% tannin. A dye extracted from the bark gives a yellow to brown colour to cloth and a red colour to leather. The branches have been used as brush fencing while the thorns have been used as needles.

A clear to red-gold gum is produced from the wood when it has been damaged by boring insects. It is collected and used as a commercial substitute for gum arabic. The gum is used in the manufacture of sweets, medicines and adhesives.

Both the pods and foliage provide forage for cattle and goats. The seeds are sometimes roasted and ground to make a coffee substitute. The sweet-smelling flowers are attractive to bees and provide excellent nectar and good pollen. Traditional medical uses of the roots include treatment of body pains, dizziness, convulsions and STDs. They are also chewed as an aphrodisiac. The roots are scattered in fowl runs to repel fowl parasites. A substance has been found in the heartwood which is said to control high blood pressure. The presence of *Acacia karroo* in grazing areas is said to indicate sweetveld and fertile soils for crops. However, it can grow on very heavy soils that become water-logged in the rainy season. It can encroach and form thickets.

Acacia karroo	
Use – Production:	firewood, timber, fodder (leaves), gum, cordage, bees, medicine
Use – Service:	soil improvement, rangeland
Growth rate:	fast
Coppicing ability:	good
Environmental tolerances:	drought, frost, salinity, waterlogging
Compatibility with arable agriculture:	moderate
Soil type:	black clay, loam or sandy
Soil fertility improvement:	very good

ACACIA NIGRESCENS

COMMON NAMES *Katopa*, *mupumbu*, *umhlope*, *umkayamhlope* (Ndebele); *mukotonga*, *mukuu*, *chinanga*, *nkoho* (Shona); *gakuanga*, *mpozva*, *sinanga* (Shangaan); *moba*, *mwaba*, *mukotokoto* (Tonga); *knobthorn* (English); *knoppiesdoring* (Afrikaans)

DESCRIPTION

SIZE/SHAPE *Acacia nigrescens* is a medium-sized erect tree up to 16 metres in height. It is leafless for much of the dry season.

BARK Dark brown to blackish, and rough, with hooked thorns on raised knobs.

TWIGS/THORNS Younger stems and branches often have large woody knobs scattered along them with small hooked thorns on top. The paired thorns are strong, hooked and blackish.

LEAVES Medium-sized, with 2-4 pairs of pinnae and normally 1-2 pairs of large round blue-green leaflets.

FLOWERS The flowers appear early in the season in August/September well before the leaves. They form long creamy white spikes.

PODS Medium in size, pointed at the end and rather papery. Dark brown in colour, they ripen and open on the tree from December to March.

SEEDS Medium-sized (6-8 x 6-8 mm) and compressed.

FIELD CHARACTERISTICS Medium to large erect tree with upright stem; strong hooked thorns on raised knobs on the trunk (not on very old trees); large leaflets. This species is easily confused with *Acacia goetzei*, but *Acacia nigrescens* has only 2-4 pairs of pinnae, each with 1 or 2 pairs of much larger leaflets.

DISTRIBUTION & ECOLOGY This tropical species is found from Tanzania southwards to KwaZulu-Natal and west to the Caprivi Strip in Namibia. It is principally a species of woodland and bushland, commonly near rivers and drainage lines. *Acacia nigrescens* requires moderately nutrient-rich but well-drained soils, and is rarely found on sandy soils. It sometimes forms woodland with mopane. It is widespread at medium and lower altitudes, generally below 1200 metres, and it is very common on clayey or rocky soils in parts of the Zambezi and Limpopo Valleys and south-east lowveld. Individuals along rivers can become very large.

GROWTH & CULTIVATION *Acacia nigrescens* is slow growing. It is tolerant of drought and fire but is quite sensitive to frost. It coppices well when young.

Acacia nigrescens

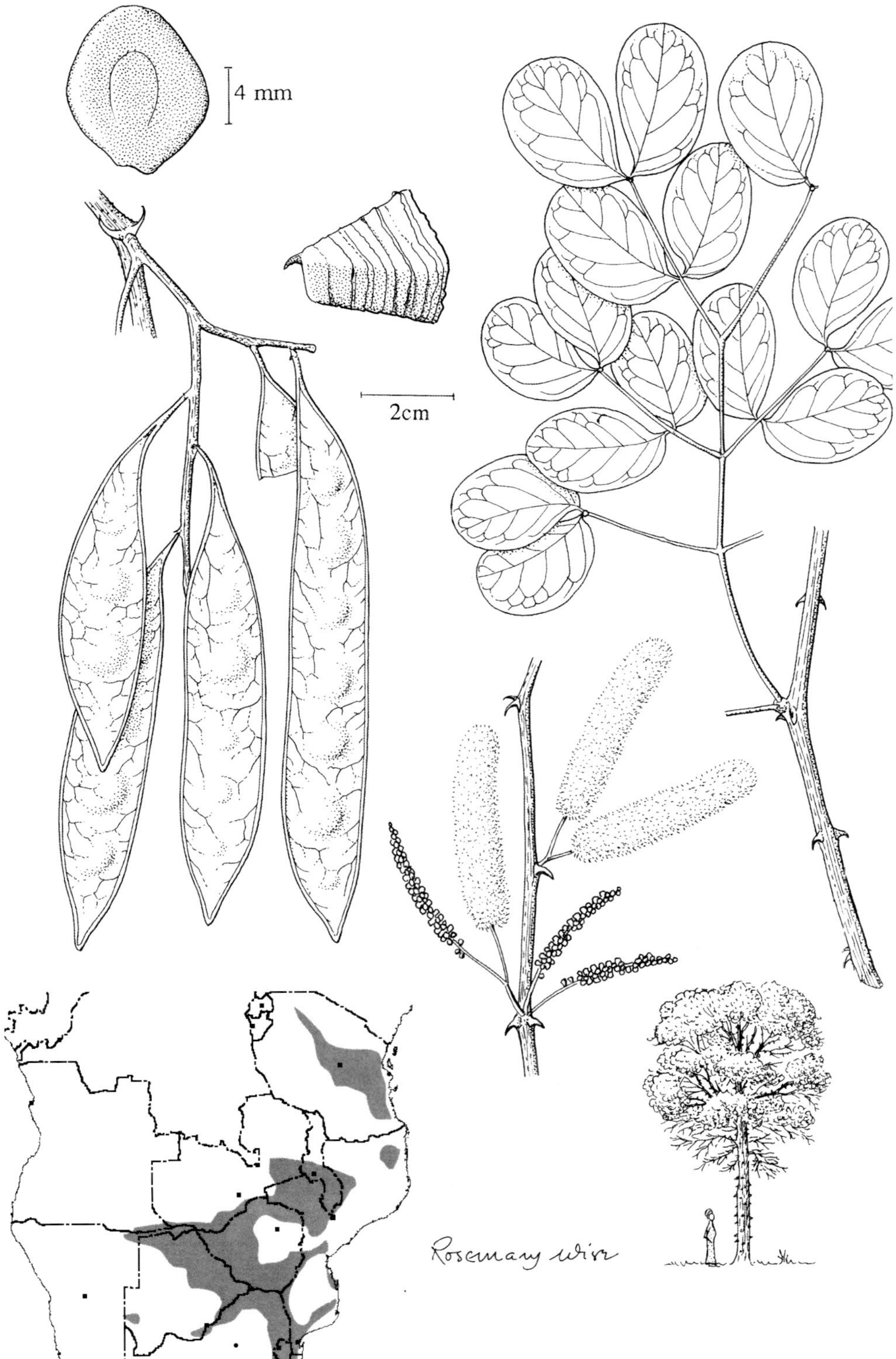

The trees produce large numbers of seeds and there is reason able germination without seed pre-treatment. The root system is generally shallow but with a deep taproot. The seedling should not be kept for long in a pot because of the long tap root, which will either start to coil up inside, or find its way rapidly into the soil beneath. Seedlings respond well to watering during dry spells. The trees can be long-lived.

USES

The trunk provides a strong durable timber which is suitable for rough construction such as hut poles, fences and kraals.The heavy, dark brown heartwood is very strong and durable but tends to crack and is difficult to work, so it is not suitable for furniture. The timber is resistant to borer, termite and fungal attack. It has been used for carving, parquet flooring, railway sleepers and mine props, and provides excellent firewood. The bark contains 15% tannin. The leaves are a good source of browse for both livestock and wildlife, particularly giraffe. Relish is made from small pale leaves. The sapwood of dead trees can house edible borer grubs. The trees provide bees with both good nectar and pollen. They are an indication of good livestock producing areas. Traditional healers make an ointment from the burnt roots to treat convulsions.

ACACIA NIGRESCENS	
Use – Production:	timber, posts, firewood, bees and browse (leaves)
Use – Service:	rangeland
Growth rate:	slow
Coppicing ability:	good
Environmental tolerances:	drought, fire
Compatibility with arable agriculture:	poor to moderate
Soil type:	clay, loam, shallow
Soil fertility improvement:	moderate

Acacia nilotica

Common names *Isanqawe, umtsharga* (Ndebele); *changaviha* (Shangaan); *nombe, mukoka, mungnombie* (Tonga); *mubayamhondoro, muunga* (Shona); *scented thorn* (English); *lekker ruikpeul*(Afrikaans)

Description

Size/shape *Acacia nilotica* is a small to medium-sized tree, usually 4-7 metres high with a broad, spreading, umbrella-shaped canopy. It keeps its leaves for much of the year.

Bark Dark brown to black and deeply furrowed.

Twigs/thorns The twigs are slightly zigzag in shape. The paired thorns are straight and whitish, often pointing slightly backwards.

Leaves Medium-sized with 2-11 pairs of pinnae and 7-25 pairs of small to occasionally medium-sized leaflets.

Flowers The sweet-smelling flowers form golden-yellow balls and appear near the beginning of the rains in October, with a second flush in January.

Pods Blackish when ripe and slightly constricted between each seed like a string of beads. They contain a sweet-smelling gum. Pods fall from the tree during the dry season without splitting open.

Seeds Medium sized (6-7 x 4.5-5 mm), fat and rounded with a hard seed coat. They are often attacked in the pod by bruchid beetles.

Field characteristics Medium-sized spreading tree; compressed dense crown; thorns straight, paired and pointing very slightly backwards; black, sweet-smelling indehiscent pods constricted between the seeds.

Distribution & ecology Various forms of this species are found all over the warmer parts of Africa to India, but there is only one form in our region. This occurs in south-western Angola and northern Namibia and also in eastern Botswana across Zimbabwe and southern Zambia to northern South Africa, KwaZulu-Natal, Mozambique and Tanzania. *Acacia nilotica* is found in a wide range of habitats, and is particularly associated with old lands and grazing areas. It appears to be mainly distributed by livestock. It is rarely found above 1500 m altitude. It usually occurs as scattered individuals rather than in stands, except around old kraals or village sites and on enriched soils such as those associated with termite mounds, drainage lines, roadsides

Acacia nilotica

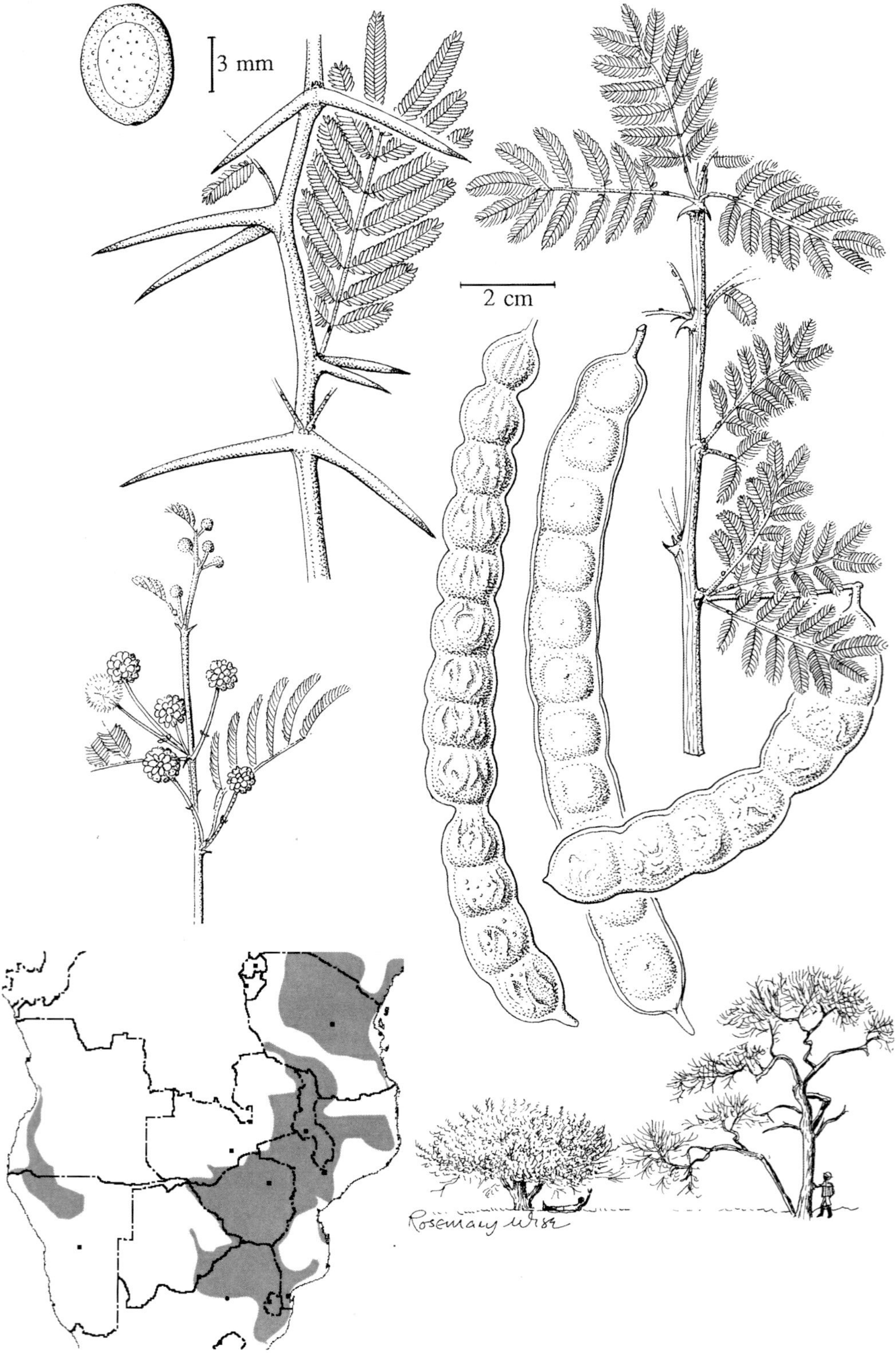

and at the foot of rocky slopes. It is also found as a stunted shrub on black cracking clay soils.

GROWTH & CULTIVATION

Acacia nilotica is slow growing and is thought to be moderately long-lived. It establishes readily from seed, although scarification is normally required, particularly if the seed is not fresh. Young plants do not compete well with grasses or weeds. Young trees coppice well. The species will tolerate only light frost, but can withstand most droughts owing to its strong deep roots. The trees are moderately tolerant of saline soils. It is reported to be possible to propagate *Acacia nilotica* from truncheons, root suckers and cuttings.

USES

Acacia nilotica has two main usable products, these are the wood and the pods. The pink-red heartwood is heavy, hard and difficult to work, so is not much used in carpentry. It is similar in hardness to *Acacia nigrescens* and *Acacia erioloba*. It seasons well, is very durable, and is resistant to borers and termites. Although straight poles are not common, it is used for fence posts and mine timber. It has also been used for tool handles and carts, and makes good firewood.

Both bark and pods are rich in tannin and have been used when tanning leather. The pods are rich in protein, as are the leaves, and livestock are often seen under trees eating the fallen pods. The tannin content probably limits intake and digestibility. In a good season an average of 35 kilograms of pods can be obtained from a mature tree.

An edible gum that can be used in confectionery and as an adhesive is obtained from the stem, and the roasted pods have been used as flavouring. The leaves, roots and gum have been used as traditional medicine in Zimbabwe to relieve coughs and colds, for eye problems, against bleeding and as a stimulant, while the flower has been used in ointments for wounds. Elsewhere the species has been used to treat haemorrhages, colds, diarrhoea, scurvy, STDs and throat infections. Sap from the pod is used as an eye medicine.

The trees provide bees with both good pollen and nectar. The tree is also a useful source of shade in grazing areas as the leaves are lost late in the dry season. Cattle readily spread the seeds of *Acacia nilotica*. When an area frequented by animals feeding on the pods is left ungrazed for a period, an invasion of these trees is possible. This can lead to bush encroachment if the area is not well managed. Grass does not grow well in its shade, although cattle often congregate under broad-canopied trees.

ACACIA NILOTICA

Use – Production:	posts, firewood, tanning, fodder (pods), gum, bees, medicines
Use – Service:	shade, soil improvement
Growth rate:	slow to moderate
Coppicing ability:	good
Environmental tolerances:	salinity, drought, but susceptible to frost
Compatibility with arable agriculture:	low
Soil type:	clay, loam, sandy, shallow
Soil fertility improvement:	very good

ACACIA POLYACANTHA

COMMON NAMES *Umdwadwa*, *umpumbu* (Ndebele); *chikwiku*, *chiungadzi*, *mugone*, *mukwakwa* (Shona); *gukwe*, *kovakova* (Shangaan); *luntwele*, *mumbu* (Tonga); *white thorn* (English)

DESCRIPTION

SIZE/SHAPE *Acacia polyacantha* is a medium to large tree, up to 15 metres high but occasionally reaching 30 metres, with an open layered canopy. The tree loses its leaves early in the dry season.

BARK Pale in colour, smooth and sometimes flaking.

TWIGS/THORNS The paired thorns are large and hooked. They are sometimes found growing on the branches but are usually less frequent high in the canopy.

LEAVES Large, with 13-40 pairs of pinnae and 26-66 pairs of very small leaflets. The petiole has a large flattened oblong gland near the base.

FLOWERS Flowers form a long, thin, creamy-white spike and appear near the beginning of the rainy season from September to December, but after those of *Acacia galpinii* and *Acacia nigrescens*.

PODS Straight, flat and papery, tapering at both ends; often many on the tree. They open while they are still on the tree towards the end of the wet season.

SEEDS Medium-sized, around 8 x 7 mm, flat and almost rectangular.

FIELD CHARACTERISTICS Tall erect tree; pale bark and layered crown spreading towards the top; hooked thorns often do not persist on the branches of mature trees; large flattened petiolar gland; early deciduous; profuse pale brown pods opening on the tree.

ACACIA POLYACANTHA

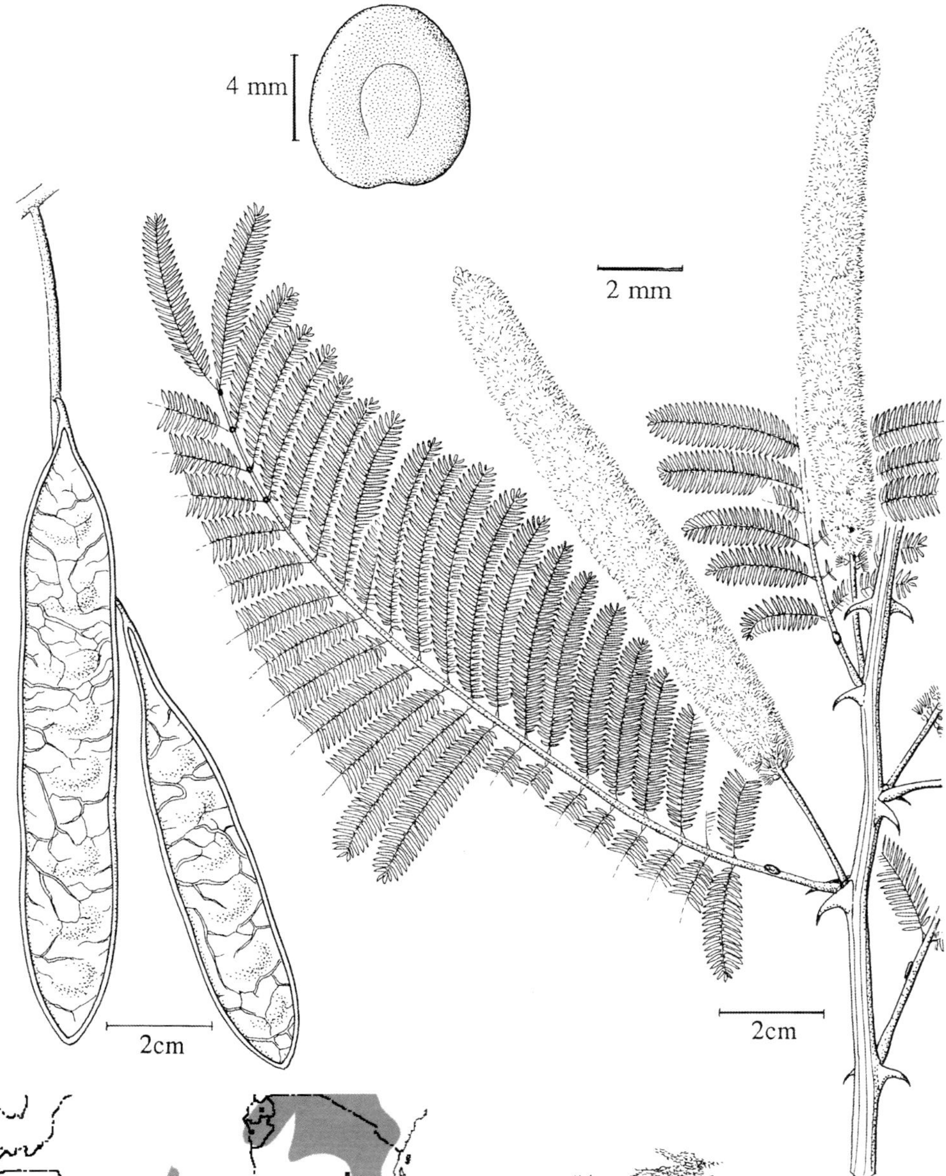

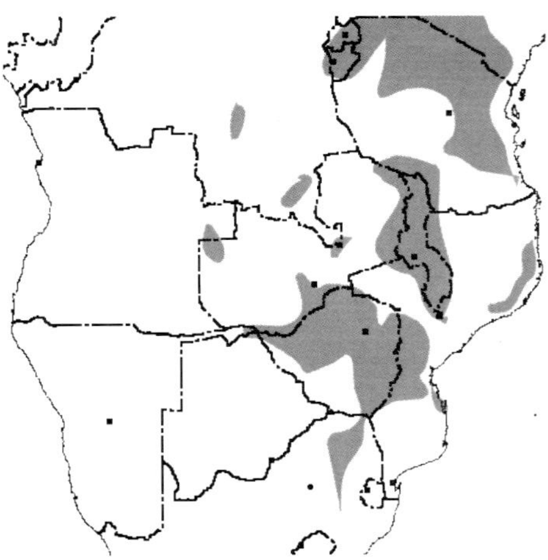

DISTRIBUTION & ECOLOGY

This essentially tropical species occurs across much of the savanna area of Africa but, in our region, only in Zambia, northern Botswana, Zimbabwe and Mozambique, with a small area in the South African lowveld. It is mostly found in open woodland and along smaller and medium-sized rivers. Its distribution appears to be limited by both frost and drought. It is particularly widespread in Zimbabwe, and is found over much of the Mashonaland highveld and Zambezi catchment, but is infrequent in the Limpopo catchment. It is generally found on red clay soils, particularly on old lands where it can be inva sive, and around wet areas and small rivers.

GROWTH & CULTIVATION

Acacia polyacantha is one of the fast-growing acacia species, but lives only for 30-40 years. The root system is shallow and spreading. The species is not drought-tolerant and loses its leaves earlier in the dry season than most acacia species. It is frost-sensitive, particularly in its early stages. Readily raised from seed, it shows good potential for breeding of improved strains. The species tends to invade old fields on clay soils, especially those on fertile red clays, where its prolific growth can become a problem. Its limited coppicing ability and small rootstock mean that *Acacia polyacantha* can be controlled by repeated cutting.

USES

The main use for *Acacia polyacantha* is timber. The red-brown heartwood has variable colouring and is moderately heavy and of moderate durability. It is prone to splitting and so needs to be seasoned slowly. The wood is good for general woodworking. Although resistant to borer and termite attack, the wood, especially the sapwood, is very susceptible to fungal attack. It has been used for crates, shelving and other indoor construction. Treated poles are used for hut construction. Although light, it makes good firewood and charcoal. The bark has been used for tanning. The fresh roots have a strong odour and have been used as a treatment for snakebite as well as a snake repellent. Bark, root and gum are also used medicinally against STDs and stomach pains. The leaves are reputed to have slight antibiotic properties. The edible gum can be used in making sweets and as an adhesive. In India, the gum has been used in dyeing and tanning. The trees provide good pollen for bees. The species is an indicator of fertile soils suitable for producing tobacco and cotton.

ACACIA POLYACANTHA	
Use – Production:	timber, firewood, tanning, gum, bees, medicine
Use – Service:	soil improvement
Growth rate:	fast
Coppicing ability:	poor

Environmental tolerances:	drought
Compatibility with arable agriculture:	poor to moderate
Soil type:	clay, loam
Soil fertility improvement:	good

ACACIA REHMANNIANA

COMMON NAMES	*Iphucula* (Ndebele); *mona* (Shona-Rusape); *mugone*, *mukwak wa* (Shona); *silky acacia* (English); *sydoring* (Afrikaans)
DESCRIPTION	
SIZE/SHAPE	*Acacia rehmanniana* is a small to medium sized tree up to 10 metres high. It branches low and has a broad rounded canopy, which remains in leaf for much of the year. It is often found as a shrub.
BARK	Orange to reddish brown on younger stems, becoming dark brown to black and furrowed on older trees.
TWIGS/THORNS	Twigs are densely covered in pale yellow to greyish hairs. The paired thorns are long, straight and greyish in colour.
LEAVES	Medium-sized, with 15-44 pairs of pinnae and numerous very small hairy leaflets.
FLOWERS	These form rather small white balls which appear from November to February.
PODS	Broad but flat and thin, with a noticeable margin. They open on the tree between April and June.
SEEDS	Medium-sized (5-9 x 5-6.5mm) and rounded in shape.
FIELD CHARACTERISTICS	Small to medium-sized tree; loose, spreading crown; rusty-red or orange stem in young trees and on branches of old trees; greyish leaves with many very small leaflets; golden-yellow hairs on young growth.
DISTRIBUTION & ECOLOGY	This species is typically associated with drier miombo or *Terminalia* woodland, and is almost confined to the Zimbabwe middle and highveld above 1000 m altitude. There are also small populations in the northern parts of the South African highveld, in southern Zambia and in Botswana around Francistown. It is common in open woodland and on vlei margins on granite and other soils, generally those with moderate clay content.

ACACIA REHMANNIANA

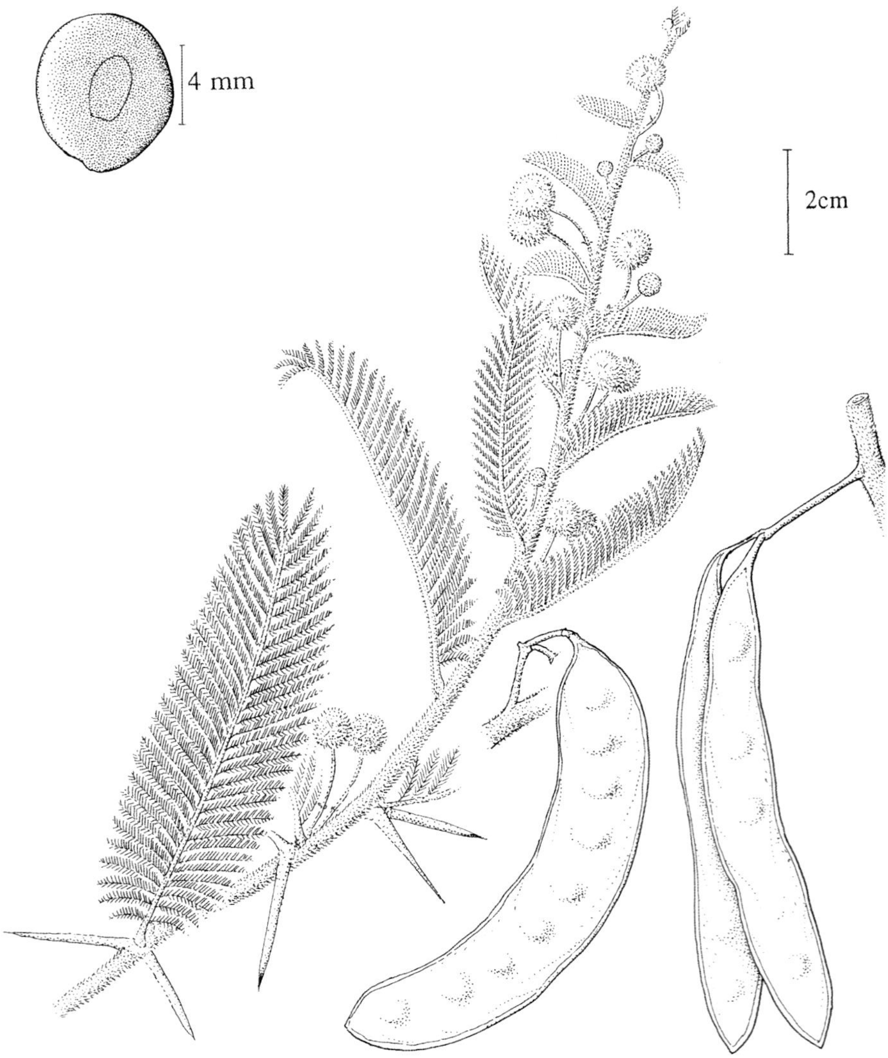

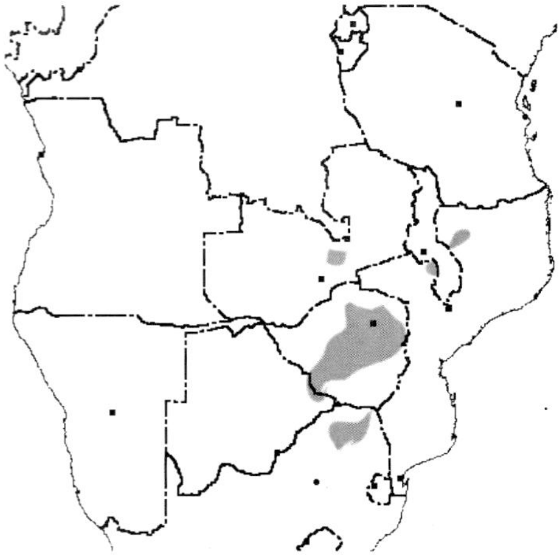

GROWTH & CULTIVATION

Acacia rehmanniana is a slow-growing species which is probably not very long-lived. It is very frost hardy except when young, and coppices well from the rootstock. The species has a tendency to become root-bound in small containers.

USES

This is not a very productive acacia. The wood is mostly sapwood and is not strong or durable as it is often attacked by borers. It does not make good firewood, mostly because of its small size. The inner bark has been used for rope. The roots are rubbed into small cuts on the body as a cure for headache and pneumonia by some traditional healers. It has been used for brush fencing, particularly when many young plants are cleared in the course of field preparation. Its growth can help rehabilitate degraded lands.

ACACIA REHMANNIANA	
Use – Production:	brush fencing, cordage, medicine
Use – Service:	soil conservation
Growth rate:	slow
Coppicing ability:	strong
Environmental tolerances:	frost, waterlogging
Compatibility with arable agriculture:	poor to moderate
Soil type:	clay
Soil fertility improvement:	moderate

ACACIA SIEBERIANA

COMMON NAMES *Umpumbu*, *umsasane* (Ndebele); *hlofungu* (Shangaan); *impangala*, *mubovwa*, *mumpangala* (Tonga); *mugone*, *mukwakwa* (Shona); *paperbark thorn*, *umbrella thorn* (English)

DESCRIPTION

SIZE/SHAPE *Acacia sieberiana* is a large tree up to 15 metres in height. It usually has a broad, flat-topped canopy, although the crown tends to become rounded in specimens found at lower altitudes. It branches low and loses its leaves early in the dry season.

BARK Light grey, rough and peeling, with a yellowish underlayer.

TWIGS/THORNS The twigs are rather slender and often slightly hairy. The paired thorns are straight and light grey. Older trees may be almost without thorns.

LEAVES Medium-sized, with 6-23 pairs of pinnae and 14-45 pairs of small-sized leaflets.

FLOWERS These form small yellow-white balls that appear early in the season between September and December, before those of *Acacia rehmanniana* or *Acacia tortilis*.

PODS Large (normally 10 x 2 cm), thick and woody, and light brown in colour. They fall off the tree late in the season without opening.

SEEDS Medium-sized to large (9-12 x 7-8 mm) and rounded in shape.

FIELD CHARACTERISTICS Large tree, generally with very wide, spreading, flattened canopy; flaking bark revealing yellowish under-layer; yellow spreading hairs on young growth; thick woody, pale brown indehiscent pods, the remains of which are often found under the tree.

The main form of *Acacia sieberiana* is found in the highveld, but a round-canopied form with narrower pods and less hairy young growth is found at lower altitudes, particularly near rivers.

DISTRIBUTION & ECOLOGY This species occurs across much of the savanna region of Africa southwards to northern Namibia, northern Botswana, Zimbabwe and Mozambique, and also in northern South Africa and KwaZulu-Natal. It is generally found on nutrient-poor sandy soils that have reasonable soil moisture levels for much of the year. In the centre of its range in Zimbabwe, *Acacia sieberiana* is found over much of the middle and high veld above 900 m altitude, generally in open areas on granite

ACACIA SIEBERIANA

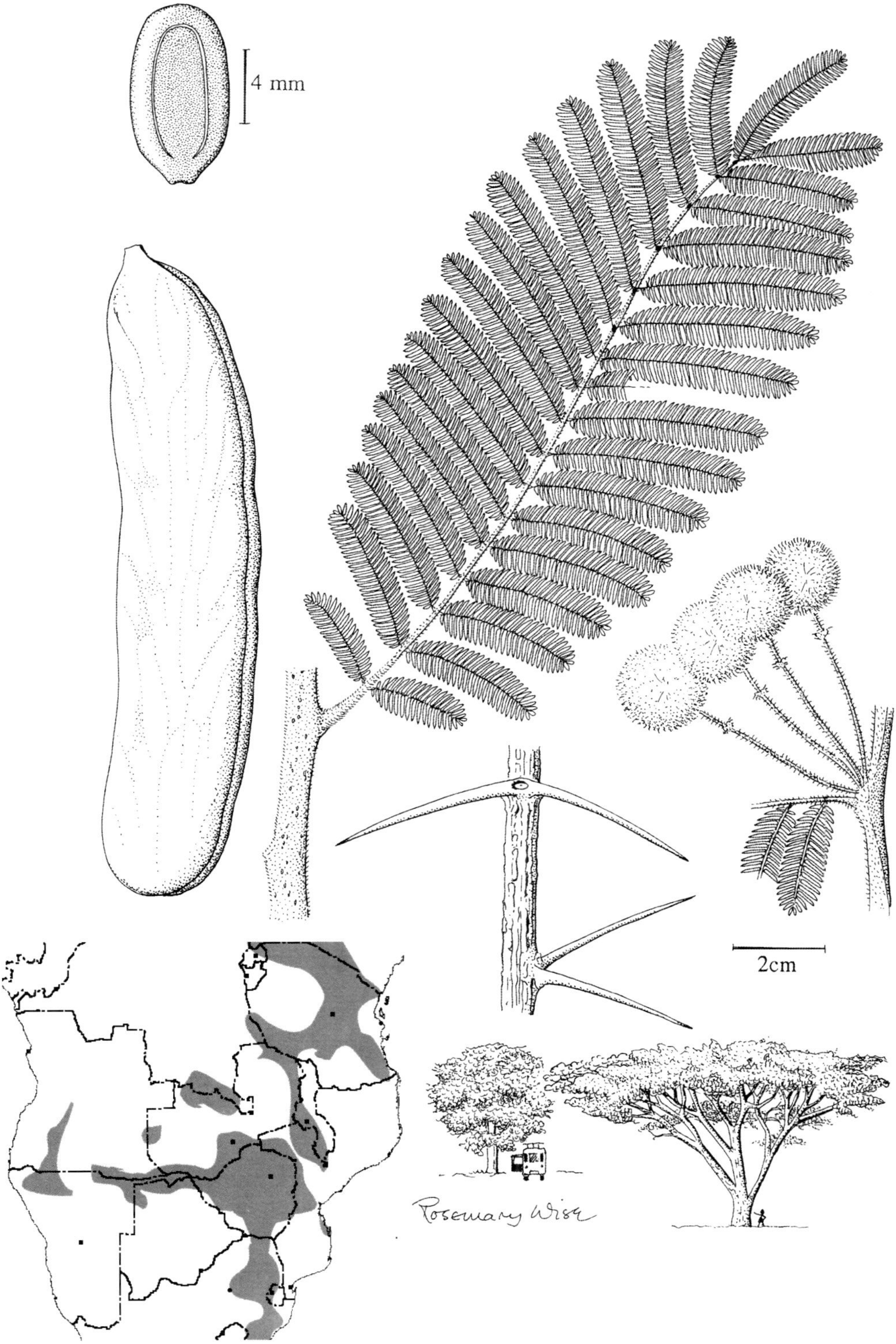

sands or along margins of vleis. Below that altitude a different form is found which is confined to areas of alluvial soil by rivers.

GROWTH & CULTIVATION *Acacia sieberiana* is a fast growing species and easy to establish from seed. It is shallow-rooted and fairly long-lived. It is both cold and frost resistant.

USES The main useful product is wood, which is mostly creamy white sapwood. This is soft and of medium density, so is not durable and needs to be well seasoned before use. The wood has been used for rough furniture and crates but it is susceptible to borers and sap stain. The trees can provide good shade. Cattle eat the pods and foliage. A single tree may produce over 200 kilograms of pods. When green, the woody pods contain a toxin which can cause fatalities in livestock. The grass under *Acacia sieberiana* trees is generally of good quality owing to nutrient input from leaf fall and partial shade. A bark infusion is used medicinally in Botswana as a painkiller and elsewhere against STDs, diarrhoea, haemorrhages and colds and as a vermifuge. In Zimbabwe, the root has been used as an antiseptic.

ACACIA SIEBERIANA	
Use – Production:	timber, fodder (pods), medicine
Use – Service:	shade, soil improvement
Growth rate:	fast
Coppicing ability:	moderate
Environmental tolerances:	frost, salinity, poor drainage
Compatibility with arable agriculture:	moderate
Soil type:	clay, sandy
Soil fertility improvement:	good

ACACIA TORTILIS

COMMON NAMES *Isanqawe*, *umsasane*, *umshishene*, *umtshatshatsha* (Ndebele); *sesani* (Shangaan); *mukoha*, *ngoka*, *muzungu* (Tonga); *muunga* (Shona); *umbrella thorn* (English); *haak-en-steek* (Afrikaans)

DESCRIPTION

SIZE/SHAPE *Acacia tortilis* is a small to large tree up to 15 metres in height. It has either a broad flat-topped canopy (subspecies *spirocarpa*) or a more rounded crown (subspecies *heteracantha*). The tree retains its leaves into the early part of the dry season.

BARK Furrowed and dark brown to blackish.

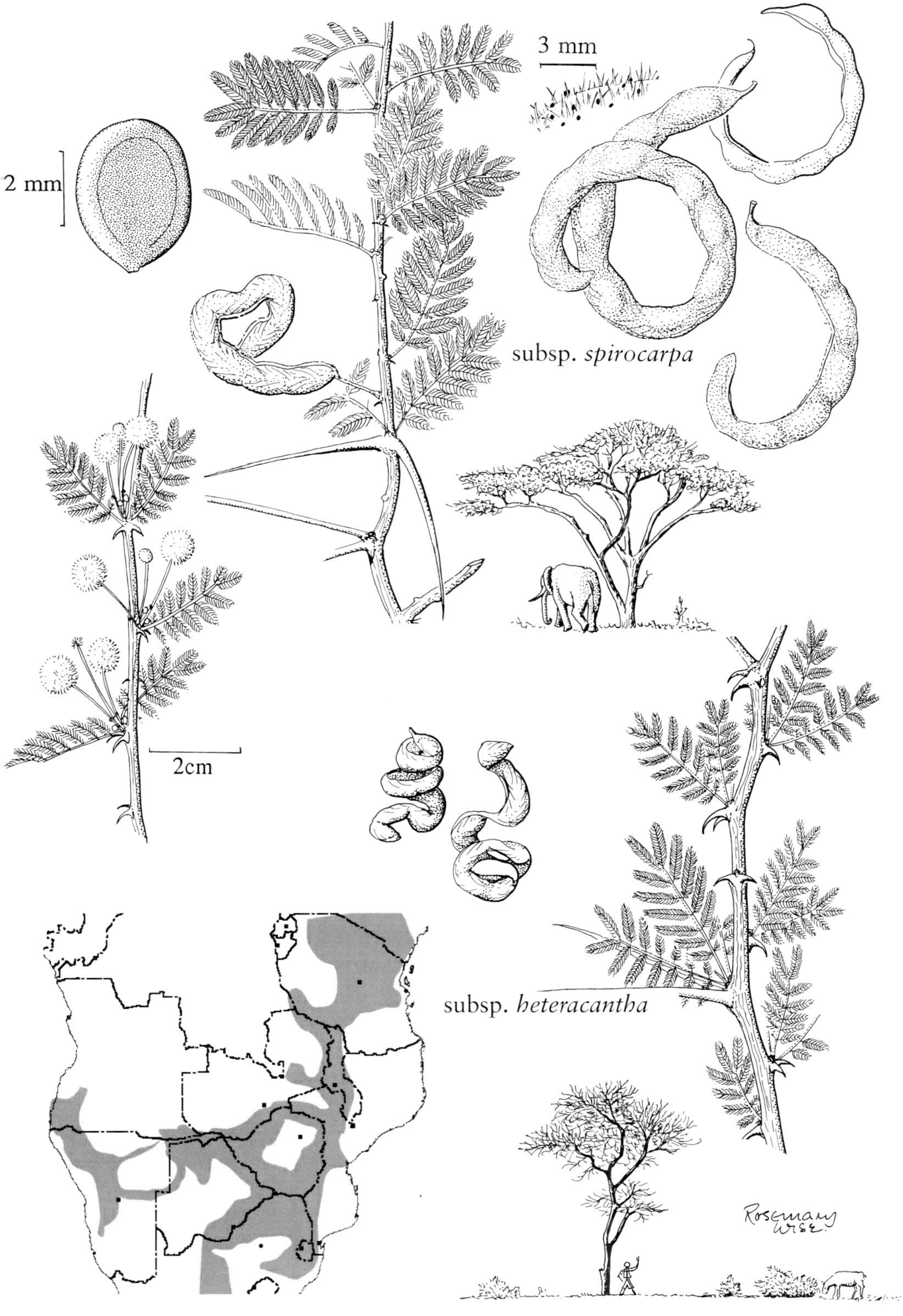
3 mm
2 mm
subsp. spirocarpa
2cm
subsp. heteracantha
Rosemary Wise

Twigs/thorns A mixture of long, straight, white paired thorns and much shorter hooked thorns are found on the twigs, sometimes a pair being formed of one of each type. Young trees and shrubs generally have many more thorns than older plants.

Leaves Small, with 2-10 pairs of pinnae and 6-19 pairs of very small leaflets.

Flowers The small balls of creamy white flowers appear from November (subsp. *heteracantha*) to April (subsp. *spirocarpa*) after rain has fallen. It is one of the last acacias to flower.

Pods Small, narrow and tightly (subspecies *heteracantha*) to loosely (subspecies *spirocarpa*) coiled.

Seeds Small (3 x 4 mm) in subspecies *heteracantha* to medium-sized (7 x 4.5-6 mm) in subspecies *spirocarpa*. The seeds are rounded and rattle inside the pods, and are often attacked by bruchid beetles.

Field characteristics Medium to large flat-topped or rounded tree; crown with flat branching to give umbrella-shaped crown (subspecies *spirocarpa*) or loosely-ascending branches to give a more rounded crown (subspecies *heteracantha*); both hooked and straight spines; minute leaflets; coiled smooth indehiscent pods.

Distribution & ecology This species is found throughout the drier parts of tropical Africa. There are two forms in our region. The more tropical form (subspecies *spirocarpa*) has hairy young twigs and only loosely coiled pods, and seems to be mostly associated with sandy alluvial soils by larger rivers below 1000 metres in altitude. This form is found from East Africa southwards to northern Namibia across to northern and eastern Zimbabwe and central Mozambique. The other form (subspecies *hetera cantha*) is generally a small to moderate sized shrub or tree with almost hairless young twigs, or when hairs are present they are less than 0.25 mm long. It has tightly coiled pods and is mostly associated with ranching areas and old lands. Subspecies *heteracantha* occurs in the Limpopo and Save catchments in Zimbabwe, southeastern Botswana, in Swaziland and in SouthAfrica down to 30 degrees south.

Growth & cultivation This pioneer species is moderate to fast growing and establishes readily on degraded lands. Subspecies *spirocarpa* is more sensitive to frost than the subspecies *heteracantha*. The tropical riverine type (subspecies *spirocarpa*) grows much faster and has moderate drought tolerance owing to its deep taproot. *Acacia tortilis* probably lives for 50-75 years. It is easily raised from seed, which remains viable for a long time.

Seedlings are intolerant of shade. However, they respond well to manure and fertiliser and to watering during dry spells. It grows well on shallow soils but must be widely spaced. Seedlings need 3-5 years of protection. The species does not intercrop well because of its wide root system. Young plants require protection from goats which otherwise keep it as a low bush as a result of their browsing.

USES

Acacia tortilis has three main usable products – wood, thorny branches and pods. The heavy wood consists mostly of yellowish sapwood, which seasons well in water. The small amount of red-brown heartwood is very heavy and hard. It makes good firewood but is not resistant enough for most construction purposes as it is readily attacked by borers. The inner bark is used to make ropes. Thorny branches are often cut and used to make an effective brush fence around arable fields to keep livestock out. The pods, especially of the non-riverine type, do not open on the tree. They contain up to 17% protein and provide a valuable source of forage for goats and cattle. The hard seeds are generally not destroyed by ingestion and are passed out in the animals' droppings. In some areas pods are gathered by farmers to feed to livestock. The leaves, which remain on the plant into the dry season, also provide browse. The flowers provide bees with a high quality pollen. The gum is edible. The roots and bark provide a yellow-brown dye. Various uses of the bark are reported in traditional medicine, including treatment of skin infections, stomach ache, diarrhoea and as a vermifuge. The species is an indicator of deep non-acid soils in good grazing areas.

ACACIA TORTILIS	
Use – Production:	firewood, fodder (leaf, pods), brush fencing, bees, cordage
Use – Service:	shade, soil improvement
Growth rate:	moderate to fast
Coppicing ability:	good
Environmental tolerances:	cold (subsp. *heteracantha*), drought, salinity
Compatibility with arable agriculture:	low to moderate
Soil type:	loam, sandy, shallow, non-acid
Soil fertility improvement:	good

DICHROSTACHYS CINEREA

COMMON NAMES *Ulugaku*, *ugagu* (Ndebele); *mupangara*, *muruka*, *musekera* (Shona); *litsenge* (Shangaan); *mugee*, *ketenge*, *wey-i*, *muweme* (Tonga); *sickle bush*, *Chinese lantern bush* (English); *sekelbos* (Afrikaans)

DESCRIPTION

SIZE/SHAPE *Dichrostachys*, which is closely related to the acacias, is a small multi-stemmed tree or shrub up to 6 metres in height. It has an untidy spreading canopy and retains its leaves for a period into the dry season. The stems are often twisted and intertwined.

BARK On larger individuals, the bark is dark grey-brown and furrowed.

TWIGS/THORNS Short spine-tipped shoots that look like thorns are found along the stem and twigs.

LEAVES Just like those of acacias. Medium-sized or large depending on the subspecies. They have 6-19 pairs of pinnae and 9-41 pairs of either small or medium-sized leaflets.

FLOWERS Consist of drooping spikes which are pink and frilly on the top portion (female) with a narrower yellow part below (male). They appear from November to January.

PODS Dark brown, clustered, twisted and convoluted. They do not break open on the plant and ripen from January to June.

SEEDS Small to medium in size (4-6 x 3-4.5 mm), hard and rounded, deep brown and glossy.

FIELD CHARACTERISTICS Shrub or small tree, often growing in disturbed areas; crown usually spiky but can become spreading and dense; short, straight single spine-tipped shoots or branches; distinctive pink and yellow flowers.

There are two forms of *Dichrostachys* in Zimbabwe. One is a shrub with relatively thin stems and smaller leaves, and is widespread on impoverished soils. The second form has larger leaves and broader leaflets with thicker stems, and is almost like a small tree. This is found in moister habitats.

DISTRIBUTION & ECOLOGY Different forms of this variable species occur across much of the savanna and dry woodland of Africa and Asia down to Australia. It is found across much of our region, although not in the arid parts of the Kalahari desert or south of the northern

DICHROSTACHYS CINEREA

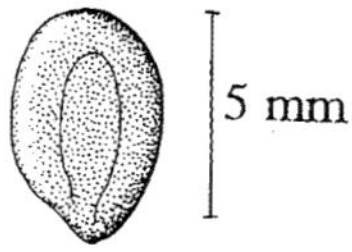

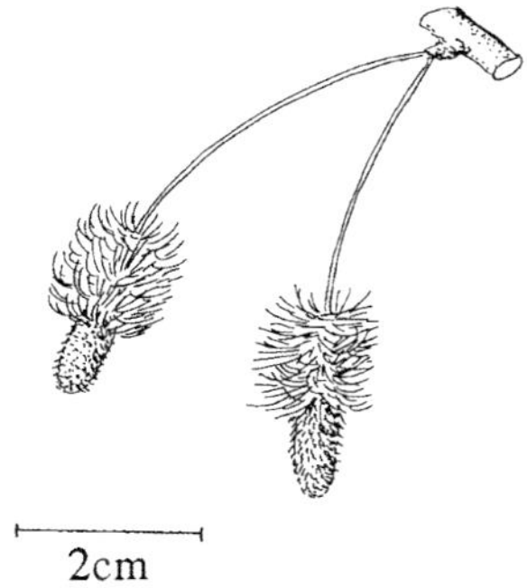

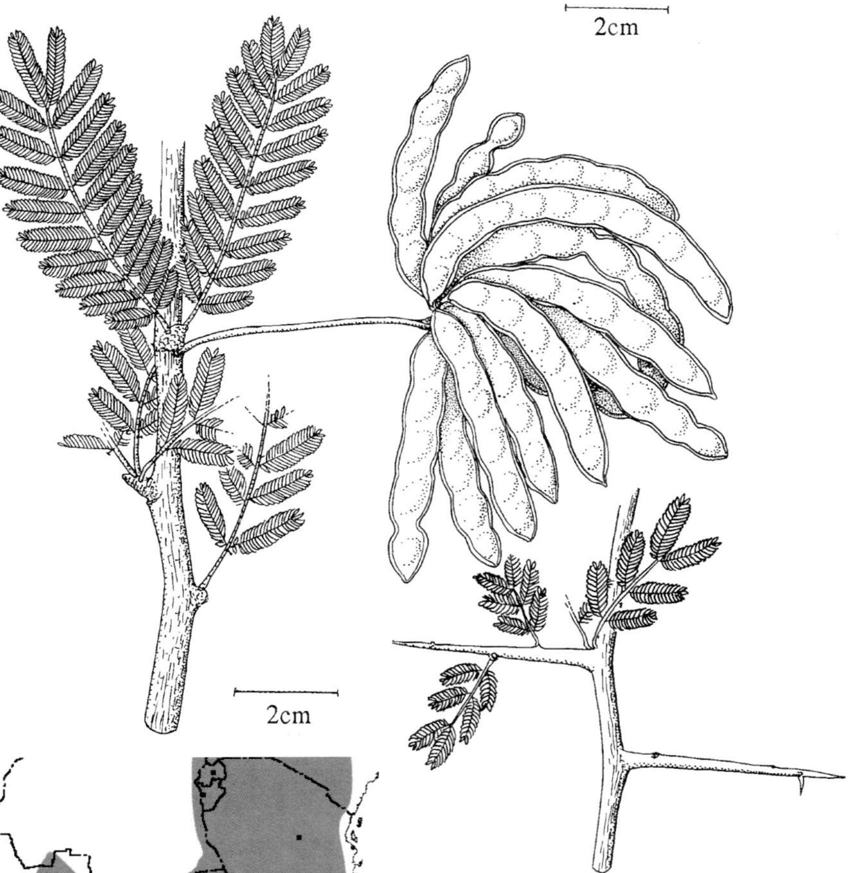

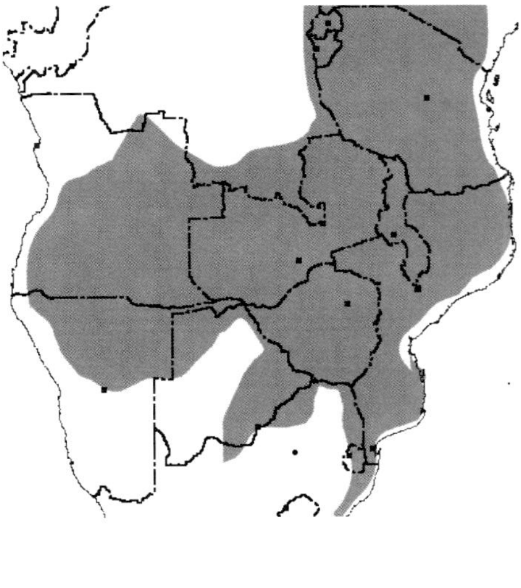

South African highveld. *Dichrostachys* occurs on a wide range of soil types ranging from Kalahari sand to cracking clay. It is particularly common in disturbed areas such as roadsides and old lands below 1500 m altitude, where it can form impenetrable thickets, and in degraded woodland. Livestock appear to be an important means of dispersal.

GROWTH & CULTIVATION

Dichrostachys is a moderately slow-growing species which readily reproduces by seed or root cuttings. It also coppices well. Thus the species is very difficult to eradicate when it has invaded an area and formed thickets. Shrubs are both cold and fire resistant. The species is tolerant of acid soils but not of saline soils. Germination is often poor and seed pre-treatment is necessary.

USES

The species produces two main products, wood and pods. The dark brown wood is strong, very heavy and hard. Owing to its generally small size, it has limited use. It is almost immune to borer, termite and fungal attack and has been used for small fence posts, hut poles, tool handles and carving. It produces excellent slow-burning firewood with good coals and is used for making charcoal. Debarked roots have been used elsewhere in Africa for strong plaiting work such as baskets and racks. The bark fibre is used to make rope. Plants are used as a live fence in some places. The pods are a valuable fodder for cattle and wildlife, but are best pulverised in a hammer mill. Leaves and shoots are also browsed. People eat the gum. The trees provide bees with a good source of nectar and pollen.

Many medicinal uses have been recorded, the majority of them utilising the root, which is used for a depressed fontanelle, as a contraceptive, against abdominal pains, influenza, cough, pneumonia, nose-bleeds, diarrhoea, infertility, backache, oedema and scorpion stings. Elsewhere in Africa, the bark has been used as a cure for dysentery, elephantiasis, worms and toothache. The leaves have been used as a diuretic, as a laxative and against boils and STDs. Chewed leaves are applied to snakebites, while the pods or seeds are also used against STDs, scabies and as a soothing eyewash. The roots and leaves can be ground and steeped in water to provide a spray against a selection of insects and nematodes. In Malawi, the leaves are used as a treatment for leprosy.

DICHROSTACHYS CINEREA	
Use – Production:	posts, firewood, cordage, browse (pods), brush fencing, medicine
Use – Service:	ornamental
Growth rate:	moderate
Coppicing ability:	good

Environmental tolerances:	cold, fire
Compatibility with arable agriculture:	poor
Soil type:	clay, loam, sand, shallow
Soil fertility improvement:	moderate

FAIDHERBIA ALBIDA

COMMON NAMES *Umpumbu*, *umtungabayeni* (Ndebele); *mutsangu* (Shona); *hlofungu*, *mukalaunga* (Shangaan); *mujagwe* (Tonga); *winterthorn*, *apple-ring acacia*, *white thorn* (English); *anaboom* (Afrikaans)

DESCRIPTION

SIZE/SHAPE *Faidherbia albida* (formerly called *Acacia albida*) is a large tree up to 25 metres high. It has a spreading canopy which is generally leafless for much of the wet season; trees come into leaf during the dry season.

BARK Whitish on younger individuals and greyish on mature trunks.

TWIGS/THORNS Zigzag in shape with straight, stout, spreading and whitish paired thorns.

LEAVES Medium-size, often curved, with 3-10 pairs of pinnae and 6-23 pairs of bluish-green medium-sized leaflets.

FLOWERS These appear around May to July, after the leaves, and consist of long, creamy white spikes.

PODS Large, fleshy and orange or brownish in colour, they are coiled like the skin of peeled fruit. The pods fall off the trees from August to November.

SEEDS Large (9-11 x 6-8 mm), rounded, with a relatively hard skin.

FIELD CHARACTERISTICS Large riverine tree with loose crown; branches long and arched in old trees; pale-cloured branchlets that zigzag between thorns towards the ends; in leaf during the dry season and leaf less and almost white in the wet season; pods orange-coloured, coiled and fleshy.

DISTRIBUTION & ECOLOGY This tropical species occurs across much of the dry tropical savanna zone of Africa and the Middle East. In our region it is mostly associated with sandy alluvium along large, seasonal ly flooded low-altitude rivers such as the Zambezi, Luangwa, Shire, Save, Pongola, Kuiseb and Limpopo, and is rarely found above 900 metres altitude. *Faidherbia* often forms extensive stands of even-sized trees on large floodplains. In West Africa it has a different distribution and ecology, being common in old lands and not confined to riverine situations.

GROWTH & CULTIVATION

Faidherbia can live to over 100 years. Growth is rapid, particularly once the taproots reach the water table. The trees have a long taproot. Rapid early growth means that trees can reach a height of 7 metres in 3 years. Young trees coppice readily. Once established, plants are tolerant of cold but not of frost. Seeds, which can be collected from August onwards, germinate readily after pre-treatment, but there is much variability in germination. For good establishment young plants should be kept free of competition from grasses. Cuttings set in sand can give up to 65% establishment. The species is tolerant of saline soils, periodic waterlogging and fire. It can grow on some slightly toxic soils, but does not do well in alkaline soils.

USES

Faidherbia provides many useful products, the main ones being wood, foliage and pods. The species has useful soil-improvement properties and is commonly used in agroforestry, especially in countries to the north of Zimbabwe. The wood is coarse-textured, soft and pale and requires seasoning to prevent splitting. In the past, large trunks were hollowed out to make dugout canoes. Despite its weakness, it has been used for light structural timber such as crates, but it is generally not suitable for outdoor application. Other uses of the wood include paddles, tool handles and rough furniture, while the wood ash is used as soap. The bark is used for beehives.

The pods and foliage are a valuable source of fodder for both livestock and wildlife, especially as both are available at the end of the dry season when there is little other available forage. The pods have a high protein content, but at times they contain a toxin when unripe. This can be removed by boiling. Large trees can produce up to 350 kg of pods in a good season, although the average pod yield per unlopped, mature tree is probably 50-160 kilograms. The seeds have been used as a famine food after boiling, removing the skins and discarding the water.

Bark, roots and crushed pods have been used as a fish poison and, in other regions, the bark has been used in traditional medicine against diarrhoea, bleeding, inflammation of the eyes and as an emetic. The flowers produce both good nectar and pollen. The trees improve soil fertility and are often interplanted with other crops. Leaf fall from the tree adds nitrogen and other nutrients to the soil and provides valuable organic material, as well as bringing up nutrients from low down in the soil profile. Competition for light and moisture is minimised as the tree comes into leaf and shows active growth only at the end of the normal cropping season.

FAIDHERBIA ALBIDA

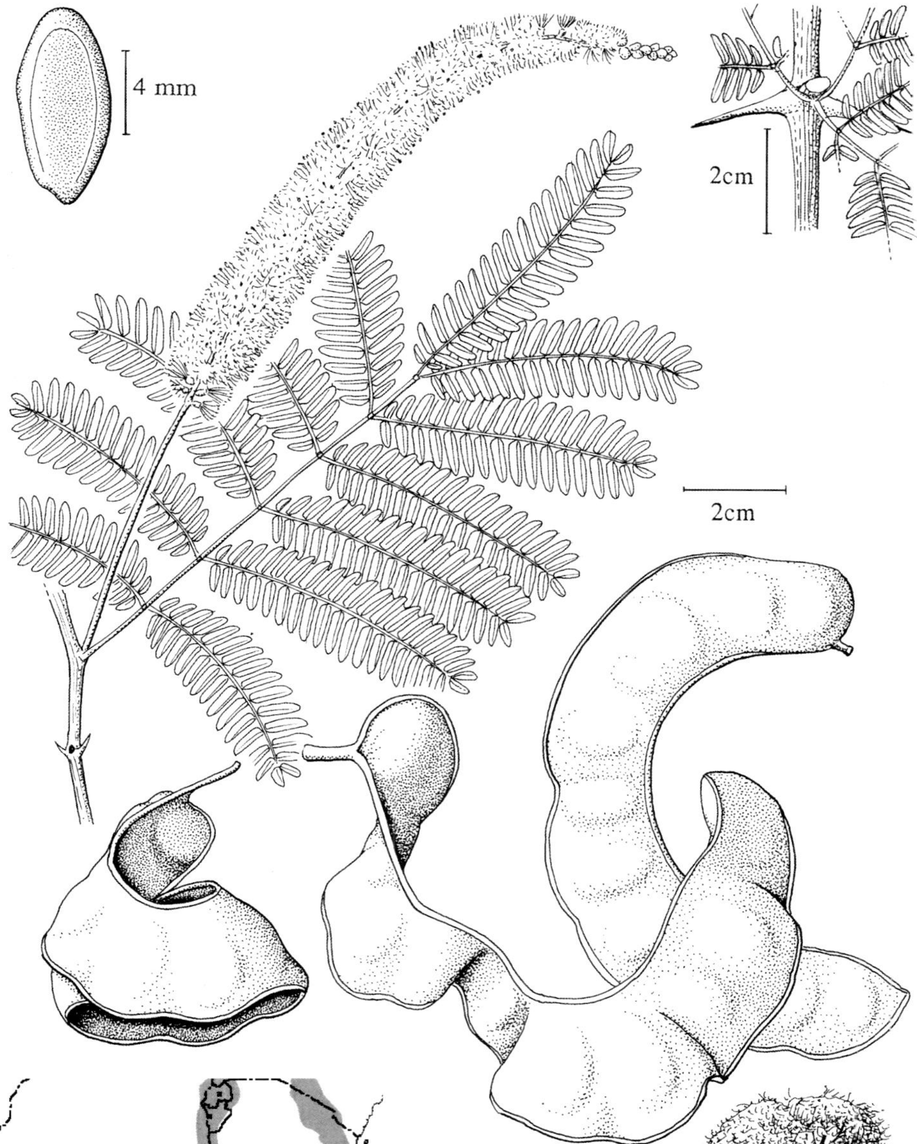

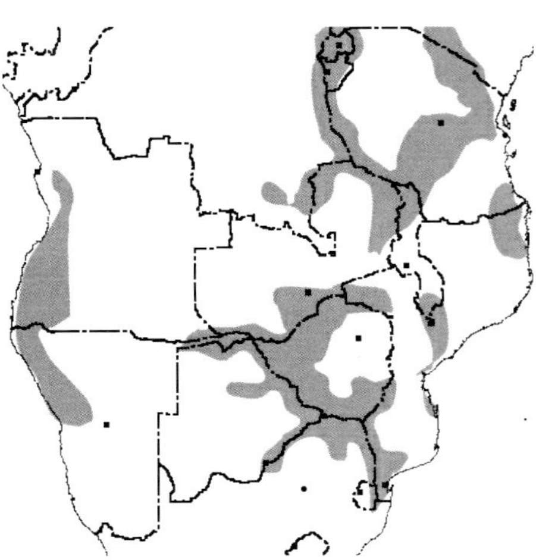

Faidherbia albida	
Use – Production:	timber, browse (leaves, pods), medicine, bees
Use – Service:	soil improvement
Growth rate:	fast
Coppicing ability:	good
Environmental tolerances:	cold, salinity, periodic (or summer) water-logging
Compatibility with arable agriculture:	good
Soil type:	sandy
Soil fertility improvement:	very good

ACACIA - THORN, FLOWER AND LEAF TYPES

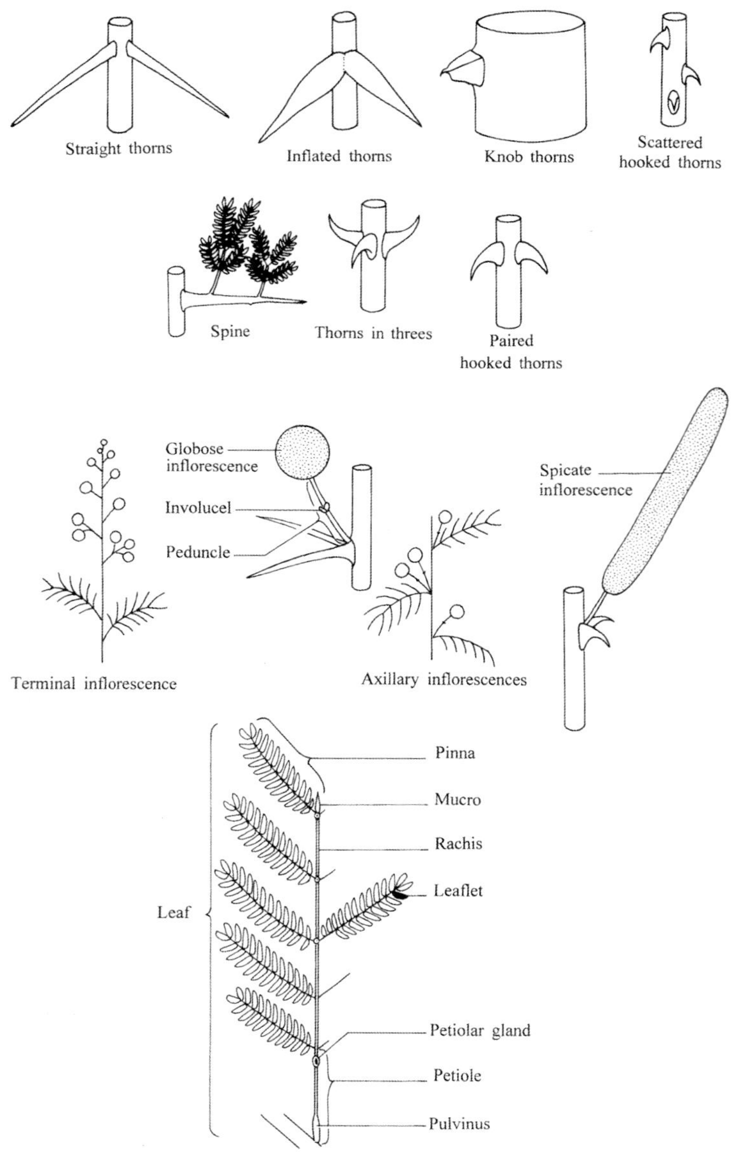

arboricide	poison for killing trees
agroforestry	growing trees in combination with crops and or livestock
alluvial	soil that has been deposited by moving water, frequently along a river bank or lake shore
ambient	surrounding
ameliorate	to make or become better; to improve
apical	at the tip or at the highest point
astringent	having a harsh, bitter quality
biodiversity	the different forms of life in a specific area
browse	green leaves, twigs and young shoots of trees or shrubs that are eaten by herbivores
bruchid	very small seed-boring beetle
calcareous	refers to soil; containing calcium carbonate, calcium or lime
canopy	the uppermost spreading branches of a tree or of a forest
colonise	become established
compound (leaf)	made up of all the leaflets, pinna and other growing parts on a single leaf stem
coppice	a tree regrowing from a sprout or a root sucker rather than from a seed; also a thicket of such trees
cotyledon	the first single leaf or one of the first pairs of leaves produced by the embryo of a flowering plant
crown	the leafy head of a tree
crude protein (CP)	the total of digestible and indigestible protein in a fodder
deciduous	having leaves that fall off annually as opposed to having leaves that remain on the tree throughout the year
degradation	reduce to a lower state
dehiscent	generally refers to pods; bursting or splitting open when ripe in order to release seed
denuded	bare, naked, having no plant or animal life remaining
dichotomous	continually dividing into two equal parts
dibble stick	a pointed tool that is used to make holes in the soil
dormant	live, but not actively growing
ecology	the science that considers the relationship between living organisms and the environment
ecozone	a habitat
elliptic	oval
embryo	the rudimentary plant contained within a seed
emetic	a medicine which causes vomiting
encroach	to invade
epidermis	the outer protective layer or skin of roots, stems and leaves
eradication	total removal
erosion	the process of wearing away
erosive	causing something, especially soil, to wear away

exotic	a plant that is not naturally occurring in a particular place but that has been introduced from elsewhere
exudate	matter which drips or oozes from pores, wounds or cuts
fissured	having long, narrow, deep cracks
flush	a sudden vigorous growth
fodder	dried browse and pods eaten by livestock
fodder bank	area set aside for growing fodder
fodder gap	time before the new flush of growth when the previous year's fodder has been used up
foliage	a mass of leaves
germinate	to sprout from a seed
gland	a small protrusion, generally on the leaf petiole, stem or pod, which secretes a fluid that is often sticky
gum	a thick, sticky liquid which is produced by some trees and shrubs that dries into a hard, brittle mass
heartwood	the hard wood at the core of a tree trunk that is often dark in colour
herbaceous	non-woody plant material
highveld	land over 1200 to 1300 metres above sea level
hilum	the point at which the seed is attached to the ovary
hydrology	the movement of water in relation to land
imbibe	to absorb or take in
impervious	incapable of being passed through or penetrated; not affected by
indehiscent	not opening at maturity to discharge seeds
indigenous	existing, growing or produced naturally in a particular place; native
inflorescence	flowers or blossoms
infusion	the liquid which results from soaking leaves, bark of some other substance in water
intolerant	unable to live under particular conditions; not able to survive
leaflet	a unit of a compound or pinnate leaf; a subdivision of a leaf
lesion	an injury; wound
live pole	a living plant that is used as a fence post; a post that takes root after it has been set in the ground
microclimate	the prevailing weather conditions in a small, distinct area
micropylar	the minute hole in the seed coat through which the pollen reaches the nucleus
monocropping	crop of a single species
mother trees	a tree from which selected seeds are sourced
nectar	the sweetish liquid in many flowers that is collected by bees to make honey
neutralise	to make ineffective
niche	the particular role of a species or organism in its community and its environment, including its position in the food cycle
nodulate	production on the roots of swellings containing bacteria

permeable	open to passage or entry, especially by fluids
petiole	the leaf stalk
pinnate	referring to a leaf which has leaflets on each side of a common axis in a featherlike arrangement
pioneer species	the first species to occupy a disturbed site, or early succession species that in their natural habitats colonise bare ground and naturally invade areas where woodland has been degraded. They have attributes that make them particularly useful for reforestation of degraded environments
pod	a dry seed vessel or fruit that splits open to release its seeds
pollard	to cut back a tree at some height above the ground so as to produce at that point a thick close growth of young branches
propagation	reproduce from the parent stock
rachis	the axis which bears the leaflets of a compound leaf
radicle	the root when it emerges from the seed
recruitment	increase in a natural population as new trees grow
rehabilitate	restore to former condition
resilience	resistance to adverse conditions
rhizobium	nitrogen fixing bacteria
ring-bark	the removal of a strip of bark from around the full circumference of a tree. This usually results in the trees death
riparian	relating to or living on the bank of a river
robustness	strength or durability
sapwood	the outer, paler and softer wood of tree
scarify	to scratch the surface of a hard seed so that it can absorb water, and thus germinate, more readily
sodic	containing a large amount of sodium as compared with calcium or potassium
spike	flowers borne on a single elongated axis
stool	the stump of a tree from which shoots spring
sucker	a subordinate shoot from a bud on the root or stem of a plant
symbiotic	association of the intimate living together of two kinds of organisms, especially when both benefit from the association
tannins	a group of astringent vegetable substances, which are often unpalatable and hinder digestion of fodder. Used in preparing leather, dyeing and the making of ink and medicine
taproot	a main root, growing almost straight downwards, from which smaller roots branch out
taxonomy	the classification of plants and other living organisms
testa	the hard, outer coat of the seed
thorn	a stiff, sharp-pointed, straight or curved woody projection on a plant
tolerant	capable of withstanding a particular environmental condition
toxin	poison
trait	a characteristic
vermicide	drug that kills worms

Acacias from different areas and habitats

Species in brackets indicate rare occurrence in this area or habitat

Areas

Highveld (1200-1500 metres)
Acacia galpinii
Acacia karroo
(Acacia nigrescens)
Acacia nilotica
Acacia polyacantha
Acacia rehmanniana
Acacia sieberiana
(Acacia tortilis)
Dichrostachys cinerea

Middleveld (750-1200 metres)
Acacia erioloba
Acacia galpinii
Acacia karroo
Acacia nigrescens
Acacia nilotica
Acacia polyacantha
Acacia rehmanniana
Acacia sieberiana
Acacia tortilis
Dichrostachys cinerea
Faidherbia albida

Zambezi valley (below 750 metres)
Acacia nigrescens
Acacia nilotica
(Acacia sieberiana)
Acacia tortilis
Dichrostachys cinerea
Faidherbia albida

Lowveld (below 750 metres)
(Acacia erioloba)
Acacia galpinii
Acacia nigrescens
Acacia nilotica
Acacia polyacantha
(Acacia sieberiana)
Acacia tortilis
Dichrostachys cinerea
Faidherbia albida

Habitats

Riverine species
Acacia galpinii
Acacia karroo
Acacia nigrescens
Acacia polyacantha
Acacia sieberiana
Acacia tortilis
Faidherbia albida
(Dichrostachys cinerea)

Vlei margins
Acacia karroo
Acacia nilotica
Acacia polyacantha
Acacia rehmanniana
Acacia sieberiana
(Dichrostachys cinerea)

Kalahari sand species
Acacia erioloba
(Acacia nilotica)
(Acacia sieberiana)
Acacia tortilis
Dichrostachys cinerea

Vernacular names of acacias from south central Africa

List of abbreviations

(Bot) Botswana; (Mal) Malawi; (Moz) Mozambique; (SA) South Africa; (Zam) Zambia; (Zim) Zimbabwe

anaboom	Afrikaans (SA)	*Faidherbia*
apiesdoring	Afrikaans (SA)	*A. galpinii*
apple-ring acacia	English	*Faidherbia*
baja	Shona (Moz)	*A. nigrescens*
butema	Kalanga (Zim)	*A. karroo*
cacana	Xitswa (Moz)	*A. tortilis*
camel thorn	English	*A. erioloba*
cananga	Shona (Moz)	*A. nigrescens*
changaviha	Shangaan (Zim)	A. *nilotica*
changua	(Gaza-Moz)	*A. nilotica*
chibombo	Bisa (Zam)	*A. polyacantha*
chigongolo	Chichewa (Mal)	*A. polyacantha*
chikwiku	Shona (Zim)	*A. polyacantha*
chimiriri	Nkhonde (Mal)	*A. tortilis*
chinese lantern bush	English	*Dichrostachys*
chipangala	Chichewa (Mal)	*Dichrostachys*
chipisyawago	Yao (Mal)	*Dichrostachys*
chisisyo	Chichewa (Mal), Shona (Moz)	*A. nilotica*
chitonya	Tumbuka, Yao (Mal)	*Faidherbia*
chiungadzi	Shona (Zim)	*A. polyacantha*
chiwiriri	Nkhonde (Mal)	*A. nilotica*
chombwe	Kunda, Soli (Zam)	*A. polyacantha*
chuanga	(Cabo Delgado-Moz)	*A. sieberiana*
condenso	Yao (Moz)	*Faidherbia*
gaba	Kalanga (Zim)	*A. karroo*
gakuanga	Shangaan (Zim)	*A. nigrescens*
ghuthu	Mbukushu (Bot)	A. *erioloba*
giraffe thorn	English	A. *erioloba*
goi	Shona (Moz)	*A. polyacantha*
gowani	Ronga (Moz)	*A. karroo*
gunga	(Sofala-Moz)	*A. sieberiana*
guvwa	Shona (Moz)	*A. polyacantha*
haak-en-steek	Afrikaans (SA)	*A. tortilis*
hibangale	Kalanga (Bot)	*Dichrostachys*
hlofungu	Shangaan (Zim)	*A. sieberiana, Faidherbia*
icuti	Macua (Moz)	*Faidherbia*
impangala	Tonga (Zim)	*A. sieberiana*
iphucula	Ndebele (Zim)	*A. rehmanniana*
isanqawe	Ndebele (Zim)	*A. tortilis, A. nilotica*
isinga	Ndebele (Zim)	*A. karroo*
kafifi	Bisa (Zambia)	*A. nilotica, A. tortilis*
kala-khanga	(Tete-Moz)	*A. polyacantha*
kalaungu	Shangaan (Zim)	*A. galpinii*
kameeldoring	Afrikaans (SA)	A. *erioloba*
kampangala	Chichewa (Mal)	*Dichrostachys*

kapele	Bemba (Zam)	*A. sieberiana*
kasale	Kaonde (Zam)	*A. polyacantha*
katenge	Kaonde, Soli, Bemba, Bisa, Tonga (Zam)	*Dichrostachys*
katopa	Ndebele (Zim)	*A. nigrescens*
knobthorn	English	*A. nigrescens*
knoppiesdoring	Afrikaans (SA)	*A. nigrescens*
kovakova	Shangaan (Zim)	*A. polyacantha*
kyambwe	Kaonde (Zam)	*A. polyacantha*
lekkerruikpeul	Afrikaans (SA)	*A. nilotica*
litsenge	Shangaan (Zim)	*Dichrostachys*
luntwele	Tonga (Zim)	*A. polyacantha*
lupangala	Bisa (Zam)	*Dichrostachys*
lusololo	Bemba, Kunda (Zam)	*Dichrostachys*
m'nyunzo	Chichewa (Mal)	*A. polyacantha*
m'sio	(Tete-Moz)	*A. nilotica*
maronga	(Niassa-Moz)	*A. sieberiana*
matenenthe	Chichewa (Mal)	*A. polyacantha*
mauexe	(Cabo Delgado-Moz)	*A. nigrescens*
mekaja	Macua (Moz)	*A. polyacantha*
messine	(Tete-Moz)	*A. nilotica*
m'sene	(Sofala-Moz)	*A. galpinii*
mfungu	Chichewa (Mal)	*A. karroo*
mfututa	Shona (Zim)	*A. galpinii*
mimosa thorn	English	*A. karroo*
minganzolo	Chichewa (Mal)	*A. sieberiana*
miwa	Macua (Moz)	*A. nilotica*
mkwanga	Yao (Mal)	*A. karroo*
mlonga	Yao (Mal)	*A. polyacantha, A. sieberiana, Faidherbia*
mnyozwa	Tumbuka (Mal)	*A. tortilis*
moba	Tonga (Zim)	*A. nigrescens*
mogabo	N.Sotho (SA)	*Faidherbia*
mogotlho	Tswana (Bot), N.Sotho (SA)	*A. erioloba*
mokala	Rolong (Bot)	*A. erioloba*
mokala	Tswana (Bot)	*A. galpinii*
mokgalo	N.Sotho (SA)	*A. nigrescens*
mokoba	Tswana (Bot)	*A. nigrescens*
mokosho	Tswana (Bot)	*Faidherbia*
moku	N.Sotho (SA)	*A. nilotica*
mologa	N.Sotho (SA)	*A. galpinii*
mona	Shona (Zim)	*A. rehmanniana*
monkey thorn	English	*A. galpinii*
moobbanziba	Tonga (Zam)	*A. nigrescens*
mooka	Tswana (Bot)	*A. karroo*
mookana	Tswana (Bot), N.Sotho (SA)	*A. karroo*
moremosetlha	Tawana (Bot)	*A. sieberiana*
moretse	N.Sotho (SA)	*Dichrostachys*
moselesele	Tswana (Bot)	*Dichrostachys*
mosu	Tswana (Bot), N.Sotho (SA)	*A. tortilis*

mosunyana	Tswana (Bot)	*A. tortilis*
motlhabakgosi	Tawana (Bot)	*A. nilotica*
motomakolotela	Mulonga (Zam)	*A. erioloba*
mpampa	Chichewa (Mal)	*A. karroo*
mpangala	Nkhonde, Tumbuka (Mal), Kalanga (Bot)	*Dichrostachys*
mpozva	Shangaan (Zim)	*A. nigrescens*
mpumbu	Kunda (Zam)	*A. sieberiana*
mtete	Yao (Mal)	*A. sieberiana*
mtubatuba	Nyanja, Soli (Zam)	*A. sieberiana*
mtzidzi	Chichewa (Mal)	*A. sieberiana*
mubaa	Lunda (Zam)	*Dichrostachys*
mubovwa	Tonga (Zim)	*A. sieberiana*
mudjwa	Mbukushu (Bot)	*A. tortilis*
mugamazi	(Gaza-Moz)	*A. nilotica*
mugee	Tonga (Zim)	*Dichrostachys*
mughandutji	Mbukushu (Bot)	*A. nigrescens*
mughombe	Mbukushu (Bot)	*A. sieberiana*
mugone	Shona (Zim)	*A. polyacantha*
mugunga	Shanjo (Zam)	*A. nigrescens*
muhoto	Lozi (Zam)	*A. erioloba*
muhwe	Ila (Zam)	*A. polyacantha*
muhwela	Tonga (Zam)	*A. polyacantha*
muina	Shona (Moz)	*A. polyacantha*
mujagwe	Tonga (Zim)	*Faidherbia*
mujwa	Shanjo (Zam)	*A. sieberiana*
mukakani	Lozi (Zam)	*A. polyacantha, Faidherbia*
mukalaunga	Shangaan (Zim)	*Faidherbia*
mukoha	Tonga (Zim)	*A. tortilis*
mukoka	Lamba, Soli (Zam)	*A. polyacantha*
mukoka	Shangaan (Zim)	*A. nilotica*
mukombwe	Lozi (Zam)	*A. galpinii*
mukone	Mbukushu (Bot)	*A. polyacantha*
mukotokoto	Tonga (Zim)	*A. nigrescens*
mukotonga	Shona (Zim)	*A. nigrescens*
mukunku	Goba, Nsenga (Zam)	*A. nigrescens*
mukuu	Shona (Zim)	*A. nigrescens*
mukwakwa	Shona (Zim)	*A. polyacantha*
mukwena	Lozi (Zam)	*A. nigrescens*
mulamba	Nkhonde (Mal)	*A. tortilis*
mumbu	Tonga (Zim)	*A. polyacantha*
mumpangala	Tonga (Zim)	*A. sieberiana*
munenje	Shona (Zim)	*A. karroo*
mungaatuwa	Tonga (Mal)	*A. sieberiana*
mungacishi	Ila (Zam)	*Faidherbia*
mungandu	Mbunda (Zam)	*A. nigrescens*
munganunshi	Bemba, Bisa (Zam)	*A. polyacantha, A. sieberiana*
mungatubetube	Bisa, Soli (Zam)	*A. sieberiana*
mungaushipa	Soli (Zam)	*A. karroo*
mungnombie	Tonga (Zim)	*A. nilotica*

mungondo	Luvale (Zam)	*Dichrostachys*
mungonga	Luchazi, Mbunda (Zam)	*Dichrostachys*
munhoto	Yao (Moz)	*A. nigrescens*
munye	Macua (Moz)	*A. nilotica*
munyenyengwe	Tonga (Zam)	*A. nilotica*
mupangara	Shona (Zim)	*Dichrostachys*
mupumbu	Ndebele (Zim)	*A. nigrescens*
muruka	Shona (Zim)	*Dichrostachys*
musangu	Tonga, Goba, Luvulu, Soli, Bisa (Zam)	*Faidherbia*
musau	Lunda (Zam)	*Faidherbia*
musekera	Shona (Zim)	*Dichrostachys*
mushehunga	Mbukushu, Mulonga (Zam)	*A. sieberiana*
mushobenka	Lozi, Mbukushu (Zam)	*A. sieberiana*
mussadzi	Shona (Moz)	*A. karroo*
musu	Lozi, Mbunda (Zam)	*A. erioloba*
muthu	Mbukushu (Zam)	*A. erioloba*
mutsangu	Shona (Zim), Sena (Moz)	*Faidherbia*
mutsangusangu	Tumbuka, Yao (Mal)	*Faidherbia*
muukila	Tonga (Zam)	*A. sieberiana*
muweme	Tonga (Zim)	*Dichrostachys*
muwina	Nkhonde (Mal)	*A. nigrescens*
muwua	Lunda (Zam)	*A. sieberiana*
muwunga	Lozi (Zam)	*A. nigrescens*
muzenze	Lunda, Luvale (Zam)	*A. sieberiana*
muzeze	Lunda (Zam)	*A. polyacantha*
muzunga	Goba (Zam), (Tete-Moz), Tonga (Zim)	*A. tortilis*
muzunganyebe	Sobu (Zam)	*A. sieberiana*
mwaba	Tonga (Zim)	*A. nigrescens*
mwanduci	Mbukushu (Zam)	*A. nigrescens*
mwehe	Lozi, Shanjo, Goba, Mbukushu (Zam)	*Dichrostachys*
n'thalala	Macua (Moz)	*Dichrostachys*
n'tshizime	Senga (Moz)	*A. sieberiana*
nachicueia	Yao (Moz)	*A. nigrescens*
nachiculia	Macua (Moz)	*A. nigrescens*
nacisungu	Ushi (Zam)	*Dichrostachys*
nakatenge	Soli, Kaonde, Ila, Tonga (Zam)	*Dichrostachys*
namalenga	Yao (Mal)	*A. nilotica, A. tortilis*
namele	Mulonga (Zam)	*A. galpinii, A nigrescens*
namuno	Macua, Yao (Moz)	*A. nigrescens*
nanyula	Yao (Mal)	*A. nigrescens*
nchongwe	Chichewa (Mal)	*A. tortilis*
ngaba	Kalanga (Bot)	*A. rehmanniana*
ngagaga	Chichewa (Mal)	*A. nilotica*
ngoka	Tonga (Zim)	*A. tortilis*
ngowe	Tumbuka (Mal), Bisa (Zam)	*A. polyacantha*
ngundanjira	Chichewa (Mal)	*A. galpinii*
nkala	Kalanga (Bot)	*A. galpinii*

nkoho	Shona (Zim)	*A. nigrescens*
nkosho	Kalanga (Bot)	*A. nigrescens*
nkotokua	Shona (Zim)	*A. galpinii*
nkungu	Chichewa, Nkhonde, Yao (Mal)	*A. galpinii, A. nigrescens*
nkwanga	Yao (Mal)	*A. sieberiana*
nombe	Shangaan (Zim)	*A. nilotica*
n'roka	Macua, Yao, Shona (Moz)	*A. nilotica, A. polyacantha, A. sieberiana*
ntalala	Macua (Moz)	*Dichrostachys*
ntarula	Nkhonde (Mal)	*A. polyacantha*
ntenge	Ronga (Moz)	*Dichrostachys*
ntkorio	Macua (Moz)	*A. nilotica*
ntonkoria	Macua (Moz)	*A. sieberiana*
nunkwekwele	Macua (Moz)	*A. nigrescens*
nyamapononđwe	Kunda (Zam)	*A. nigrescens*
nyungwe	Chichewa (Mal)	*A. polyacantha*
ocenca	Kunda (Zam)	*A. nilotica*
omue	Herero (Bot)	*Faidherbia*
omuhengehenge	Herero (Bot)	*A. sieberiana*
omumbonde	Herero (Bot)	*A. erioloba*
orusu	Herero (Bot)	*A. karroo*
orusu-orupunguya	Herero (Bot)	*A. tortilis*
pangala	(N.Moz)	*Dichrostachys*
paperbark thorn	English	*A. sieberiana*
scented thorn	English	*A. nilotica*
sekelbos	Afrikaans (SA)	*Dichrostachys*
sesani	Shangaan (Zim), Ronga (Moz)	*A. tortilis*
sickle bush	English	*Dichrostachys*
silky acacia	English	*A. rehmanniana*
sinanga	Shangaan (Zim)	*A. nigrescens*
siriri	(Tete-Moz)	*A. tortilis*
soetdoring	Afrikaans (SA)	*A. karroo*
sondelangange	Zulu (SA)	*A. nigrescens*
sweet thorn	English	*A. karroo*
sydoring	Afrikaans (SA)	*A. rehmanniana*
tchissio	Sena (Moz)	*A. nilotica*
thalala	Macua (Moz)	*A. nilotica*
tsenga	Ronga (Moz)	*Dichrostachys*
tuitui	Kunda (Zam)	*A. sieberiana*
ugagane	Zulu (SA)	*Dichrostachys*
ugagu	Ndebele (Zim)	*Dichrostachys*
ulugaku	Ndebele (Zim)	*Dichrostachys*
umbambampala	Zulu (SA)	*A. nigrescens*
umbrella thorn	English	*A. sieberiana, A. tortilis*
umdwadwa	Ndebele (Zim)	*A. galpinii, A. polyacantha*
umhlalankwazi	Zulu (SA)	*Faidherbia*
umhlope	Ndebele (Zim)	*A. nigrescens*
umholo	Ndebele (Zim)	*A. erioloba*
umkamba	Zulu (SA)	*A. sieberiana*
umkayamhlope	Ndebele (Zim)	*A. nigrescens*

umkhaya	Zulu (SA)	*A. nigrescens*
umkhaya-wemfula	Zulu (SA)	*Faidherbia*
umnqawe	Zulu (SA)	*A. nilotica*
umpumbu	Ndebele (Zim)	*A. polyacantha, A. sieberiana. Faidherbia*
umsasane	Ndebele (Zim), Zulu (SA)	*A. tortilis*
umshishene	Ndebele (Zim)	*A. tortilis*
umtshanga	Ndebele (Zim)	*A. nilotica*
umtshatshatsha	Ndebele (Zim)	*A. tortilis*
umtungabayeni	Ndebele (Zim)	*A. galpinii, Faidherbia*
umtwetwe	Zulu (SA)	*A. tortilis*
umunga	Zulu (SA)	*A. karroo*
uzimwe	Kunda (Zam)	*A. nilotica*
wey-i	Tonga (Zim)	*Dichrostachys*
white thorn	English	*A. polyacantha, Faidherbia*
winterthorn	English	*Faidherbia*
xinghave	(Gaza-Moz)	*A. nilotica*

BIBLIOGRAPHY

Astle, W.L., Phiri, P.S.M. & Prince, S.D. (1997). A dictionary of vernacular-scientific names of plants in the mid-Luangwa Valley, Zambia. *Kirkia* 16: 161-203.

Binns, B. (1972). *Dictionary of Plant Names in Malawi.* Government Printer, Zomba, Malawi.

Bingham, M.G. (1990). An ethno-botanical survey of Senanga West. Consultancy report, RDP Livestock Services for Ministry of Agriculture, Senanga District, Zambia.

Bingham, M. (2001). Personal communication. Lusaka, Zambia.

Coe, M and H. Beentje (1991). *A Field Guide to the Acacias of Kenya.* Oxford University Press, Oxford, 148 pp.

Fanshawe, D.B. & Mutimushi, J.M. (1971). A checklist of plant names in the Lozi languages. Forest Research Bulletin No. 23. Forestry Department, Kitwe, Zambia.

Koning, J. de (1993). *Checklist of vernacular plant names in Mozambique.* Wageningen Agricultural University Papers 93.2. Agricultural University, Wageningen, Netherlands.

Ross, J.H. (1971). *The Acacia Species of Natal.* Wildlife Protection and Conservation Society, Durban.

Smit, N. (1999). *Guide to the Acacias of South Africa.* Briza Publications, Pretoria, South Africa, 223 pp.

Tietema, T., Merkesdal, E. & Schroten, J. (1992). *Seed Germination of Indigenous Trees in Botswana.* ACTS Press, Nairobi, Kenya.

Timberlake, J.R. (1980). *Handbook of Botswana Acacias.* Ministry of Agriculture, Gaborone, Botswana 180 pp.

Timberlake, J.R. (1984). List of vernacular names recorded from Montepuez area, Cabo Delgado. Unpublished manuscript. INIA, Maputo, Mozambique.

Timberlake, J.R., Fagg, C. & Barnes, R. (1999). *Field Guide to the Acacias of Zimbabwe.* CBC Publishing, Harare, 160pp.

Venter, F. & Venter, J-A. (1996). *Making the Most of Indigenous Trees.* Briza Publications, Pretoria, South Africa.

Notes

Flowering, fruiting and leafing times of acacias and allied species

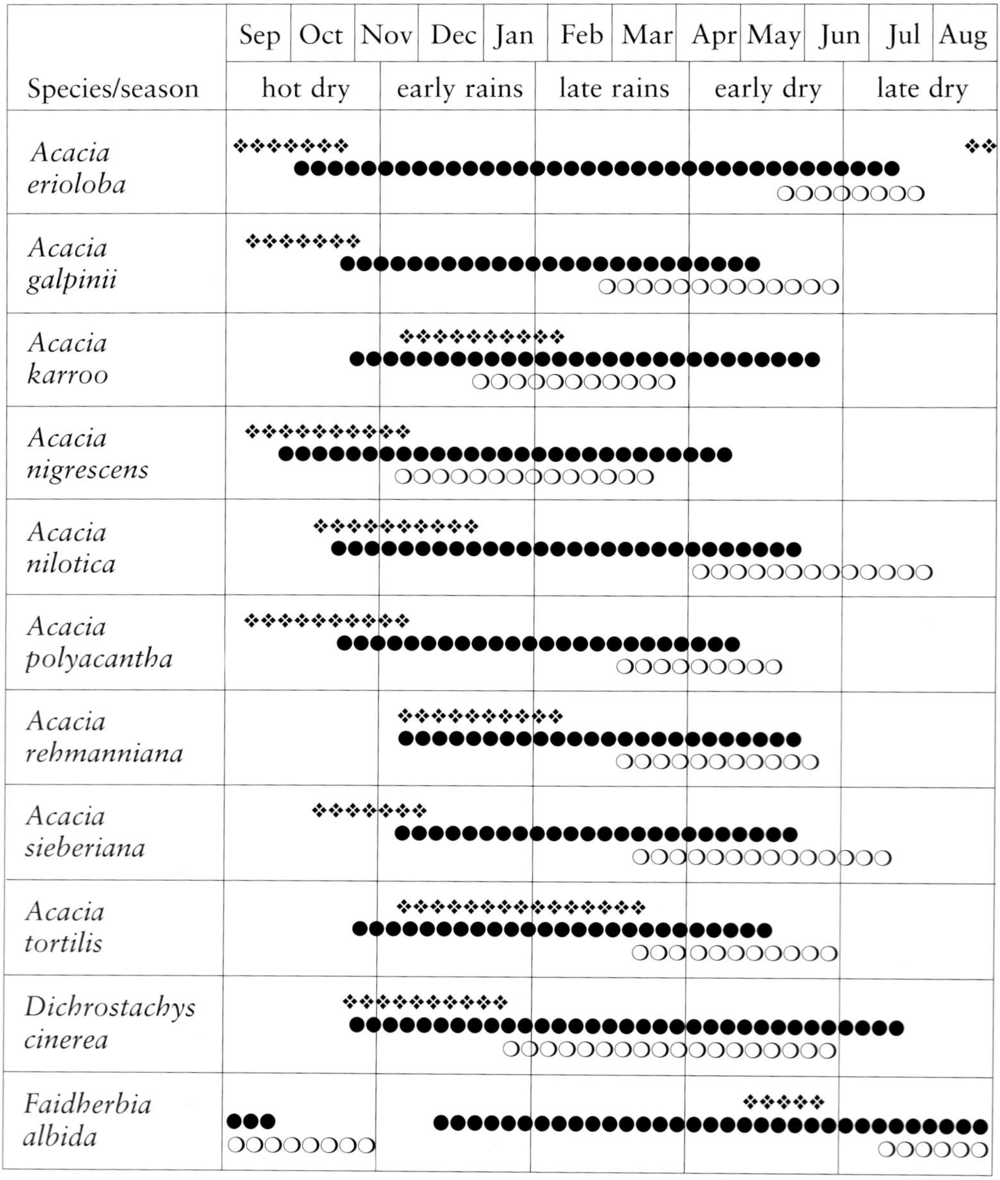

Key: ❖❖❖❖ Flowering
●●●● In leaf
○○○○ Ripe pods

1

ACACIA ERIOLOBA

Large spreading tree on sandy soils: zigzag shape of young twigs; large straight thorns often with swollen bases; **large woody, grey, velvety ear-shaped pods**; dark grey-green foliage.

2

ACACIA GALPINII

Tall, magnificent tree usually found on riverbanks or termite mounds; big, loose crown when mature; sometimes with a flaky bark when young; **long thin spikes of cream flowers with reddish tinge before the leaves come out.**

3

ACACIA KARROO

Medium-sized, upright tree along watercourses and on black clay soils; rounded crown with curved ascending branches; bark almost black on trunk, rusty brown or orange on branches; glossy, rich-green leaflets; bunches of slender sickle-shaped pods. **Profuse, golden-yellow balls of flowers from November to March.**

4

ACACIA NIGRESCENS

Medium to large erect tree with upright stem; **strong hooked thorns on raised knobs on the trunk** (not on very old trees); large leaflets.

5

ACACIA NILOTICA

Medium-sized spreading tree; compressed dense crown; thorns straight, paired and pointing very slightly backwards; **black, sweet-smelling, indehiscent pods constricted between the seeds.**

6

ACACIA POLYACANTHA

Tall erect tree; with pale bark and **layered crown spreading towards the top**; hooked thorns often do not persist on the branches of mature trees; large flattened petiolar gland; early deciduous; profuse pale brown pods opening on the tree.

7

ACACIA REHMANNIANA

Small to medium-sized tree; loose, spreading crown; **rusty-red or orange stem in young trees and on branches of old trees**; greyish leaves with many very small leaflets; golden-yellow hairs on young growth.

8 **Acacia sieberiana**

Large tree, generally with very **wide spreading flattened canopy**; flaking bark revealing yellowish under-layer; yellow spreading hairs on young growth; thick woody, pale brown indehiscent pods, the remains of which are often found under the tree.

9 **Acacia tortilis**

Medium to large flat-topped or rounded tree; crown with flat branching to give umbrella-shaped crown (subspecies *spirocarpa*) or loosely ascending branches to give a more rounded crown (subspecies *heteracantha*). Both hooked and straight spines; minute leaflets; **coiled smooth indehiscent pods.**

10 **Dichrostachys cinerea**

Shrub or small tree, often growing in disturbed areas; crown usually spiky but can become spreading and dense; short, straight, single spine-tipped shoots or branches; **distinctive pink and yellow flowers;** a convoluted 'ball' of dark brown pods.

11 **Faidherbia albida**

Large riverine tree with loose crown; branches long and arched in old trees; **pale-coloured branchlets that zigzag between thorns towards the ends;** in leaf during the dry season and leafless and almost white in the wet season; pods orange-coloured, coiled and fleshy.